STRIKING THE
HORNETS'
NEST

STRIKING THE
HORNETS'
NEST

NAVAL AVIATION AND THE ORIGINS OF
STRATEGIC BOMBING IN WORLD WAR I

GEOFFREY L. ROSSANO
THOMAS WILDENBERG

Naval Institute Press
Annapolis, Maryland

This book has been brought to publication with the
generous assistance of Marguerite and Gerry Lenfest.

Naval Institute Press
291 Wood Road
Annapolis, MD 21402

Library of Congress Cataloging-in-Publication Data
Rossano, Geoffrey Louis.
  Striking the hornets' nest : naval aviation and the origins of strategic bombing in
World War I / Geoffrey L. Rossano and Thomas Wildenberg.
    pages cm
  Includes bibliographical references and index.
  ISBN 978-1-61251-390-4 (alk. paper) — ISBN 978-1-61251-391-1 (ebook) 1.
World War, 1914-1918—Aerial operations. 2. Naval aviation—History—20th cen-
tury. 3. Bombing, Aerial—History—20th century. 4. Great Britain. Royal Naval Air
Service—History. 5. United States. Navy—Northern Bombing Group—History.
6. Air power—History—20th century. I. Wildenberg, Thomas, 1947- II. Title. III.
Title: Naval aviation and the origins of strategic bombing in World War I.
  D602.R67 2015
  940.4'4—dc23

(∞) Print editions meet the requirements of ANSI/NISO z39.48–1992 (Permanence
of Paper).

Printed in the United States of America.

23 22 21 20 19 18 17 16 15   9 8 7 6 5 4 3 2 1
First printing

To all Naval Aviators past, present, and future,
who place themselves in harm's way to defend the
United States of America

The most ambitious operational project
undertaken by Naval Aviation during World War I.

—Clifford Lord

Contents

Acknowledgments

C ompleting a project such as this depends extensively on the enthusiastic assistance of many individuals and institutions. The authors
have relied particularly on the collections and staff of several major
repositories, including the National Archives, the Library of Congress, the
Naval History and Heritage Command, and the Archives and Special Collections Branch of the Library of the Marine Corps. We would like to offer
special thanks to Joshua Stoff of the Cradle of Aviation Museum of Garden
City, New York; Pati Threatt, Archivist and Special Collections Librarian,
Frazar Memorial Library, McNeese State University, Lake Charles, Louisi-
ana; Mike Miller, Greg Cina, and Jim Ginther of the Archives and Special
Collections Branch of the Library of the Marine Corps; now-retired Air
Force Historian Roger G. Miller; and Barbara Gilbert, Archivist, Fleet Air
Arm Museum. Nattalie Will graciously prepared the maps for this volume.
Finally, we wish to acknowledge our appreciation of the work of the Naval
Institute Press and its director, Rick Russell.

Abbreviations

FDRPLM	Franklin D. Roosevelt Presidential Library and Museum, Hyde Park, N.Y.
LC	Library of Congress, Washington, D.C.
NA	National Archives, Washington, D.C.
NHHC	Naval History and Heritage Command, Washington, D.C.
MCL	Marine Corps Library and Archives, Quantico, Va.
OpNav	Office of the Chief of Naval Operations
RG	Record Group
SecNav	Secretary of the Navy
SHM	Service Historique de la Marine, Paris

STRIKING THE
HORNETS'
NEST

Introduction

Worldld War I witnessed military conflict on a previously
unimaginable scale. Older strategies such as trench warfare and maritime blockade were expanded, joined by
revolutionary new concepts and technologies that included the first
widespread use of poison gas, radio, gasoline-powered transport, dreadnoughts, tanks, airplanes, and submarines. During the war, United
States naval aviation participated directly and aggressively in this seis-
mic change. The largest of the Navy's aeronautic activities, ultimately named
the Northern Bombing Group (NBG), represented a pathbreaking
attempt to implement an innovative concept with far-reaching implications known as strategic bombing. The program envisioned the use
of scores, then hundreds, of day and night bombers to destroy German U-boat facilities located along the Belgian coast, something President Woodrow Wilson described as striking the hornets in their nest.
Efforts to employ long-range aircraft to conduct strategic missions also
extended to the establishment of a bombing base at Killingholme, England, development of speedy "sea sleds" to launch heavy bombers from
the sea, and plans to create a Southern Bombing Program to attack
Austrian military infrastructure in the Adriatic region.

Throughout World War I, the greater part of the Allies' air strength
provided tactical support to the land armies. Nearly all the strategic
bombing operations conducted on the Western Front before the formation of the Royal Air Force (RAF) in 1918, were undertaken by the Royal
Naval Air Service (RNAS). The French preferred not to carry out such
missions for fear of instigating German reprisals. And while the Italian
air service did conduct a number of strategic raids, tactical considerations
remained its principal preoccupation. Because early strategic bombing

was practiced on a very small scale, its direct impact on the future development of this form of aerial warfare has been overlooked by most historians. Thus, little has been written about the important role played by the United States Navy's Northern Bombing Group or its mentor and patron the Royal Naval Air Service. This omission needs redressing.

In the early months of the war, the RNAS carried out its first raids against targets in Germany, and by the end of 1915 plans were in hand for a systematic offensive against enemy industrial centers. From 1916 onward, and especially during the last year of the war, much work, involving practical experiments, operation research, and staff planning, was undertaken in the field and a considerable body of knowledge assembled. As Neville Jones related in his seminal work *The Origins of Strategic Bombing,* "The leaders of the Royal Naval Air Service were convinced that strategic bombing had a valuable part to play in the air war and made great efforts to provide suitable aircraft and equipment for this work."[1] But their endeavors were frustrated by a scarcity of resources and the aerial demands of the land battle that consumed pilots and planes at a prodigious rate. In desperation, they turned to the United States Navy and Marine Corps.

The United States Navy, like its British counterpart, had traditionally played a wide-ranging strategic role with the emphasis on mobility and flexibility of response, and a few visionaries grasped the opportunity offered by the airplane as an offensive weapon. But the Navy, having arrived late on the European scene, depended utterly on the cooperation, expertise, guidance, facilities, and equipment of its allies. The large strategic bombing offensive eventually planned against enemy submarine facilities followed a blueprint created by the Royal Naval Air Service, and revealed the Americans' deference toward British priorities and their reliance upon the assistance of their allies.

Royal Navy officers such as Captain Charles Lavorock Lambe and Commander Spenser Grey worked ceaselessly to instruct, inspire, and support their new protégés. The partnership proved mutually beneficial. The Americans pursued a mission the British could not, while receiving indispensable logistical and training support. The U.S. Navy's offensive campaign in Flanders occurred in isolation from its other activities in France, Ireland, and Italy.

The intimate cooperation of United States and British naval air services in pursuing the goal of establishing a round-the-clock strategic assault against enemy submarine infrastructure extended to the point of Americans operating under RNAS/RAF command, even assigning Navy personnel in large numbers to British squadrons. Such actions reflected the cooperative approach of Adm. William S. Sims, the commander of U.S. naval forces operating in Europe (headquarters in London), and Capt. Hutchinson I. "Hutch" Cone, the head of U.S. naval aviation in the region (headquarters in Paris).

Though many in Washington, including President Woodrow Wilson, Secretary of the Navy Josephus Daniels, and Chief of Naval Operations (CNO) Rear Adm. William S. Benson, held a deep-seated distrust of the British, senior commanders in London and Paris suffered from no such misgivings. Even as Gen. John Pershing, commander of the American Expeditionary Force (AEF), resisted amalgamation of his troops with a "backs-to-the-wall" tenacity, Sims and Cone embraced cooperation, even subservience, as the most logical method to achieve their goals of establishing a credible naval aviation force in Europe and launching a crippling assault on the enemy's U-boat efforts to cut the Atlantic life line.

The eventual size and scope of the Northern Bombing Group provide an important lens through which to analyze naval aviation's first major military campaign. In 1917–18, strategic bombing stood far outside the Navy's previous aeronautic plans, experience, or expectations of fleet-related activities, but nonetheless reflected both the very aggressive anti-submarine doctrine of the Woodrow Wilson administration and the Navy Department, and American dissatisfaction with the Royal Navy's efforts in this field. In fact, despite encountering numerous difficulties throughout its life, the Northern Bombing Group experience inspired calls from Washington, London, and Paris during the summer and fall of 1918 for more, not less, bombing.

Historians Stephen Harris and Robin Higham noted recently that "Air Forces are material organizations utterly dependent on complex understructures."[2] They are thus dominated by issues of technology, production, and logistics. Forces must be recruited, trained, equipped, and transported, as well as operate in challenging conditions. The Northern Bombing Group embodied this paradigm and faced each of these challenges, the most difficult being acquisition of sufficient aircraft capable of conducting the planned mission. In fact, the program depended completely on securing these machines, which proved to be the Achilles' heel of the entire effort. The Allies simply did not have enough aircraft to share and the United States could not develop, build, or transport its own to the battlefield in time.

The Navy's unexpected embrace of a strategic bombing mission resulted from a unique mix of events and personalities. It drew urgency from the mortal threat posed by the German submarine offensive and the aggressive stance of President Wilson and the Navy Department hierarchy, coupled with their persistent belief that the Royal Navy either could not or would not employ a more aggressive strategy against the U-boat. The French obsession with Dunkirk and the related early activities of Lt. Kenneth Whiting in Europe pointed the Navy toward the land-sea battlefront in Flanders and the work carried out there by the Royal Naval Air Service, while also offering a blueprint for future action. In an attempt to gain wider support in the Navy Department, aviators never ceased trumpeting the virtues of their new specialty. The enthusiasm of Admiral Sims and Captain Cone for an aerial offensive lent critical support, as did the work of the Planning Section in London and the General Board in Washington. The willingness of CNO Benson and Secretary of the Navy Daniels to endorse such programs turned the tide. Lacking any single key piece, the program never could have existed.

The range of offensive aviation initiatives—proposed, planned, and implemented—at least partially rebuts dated notions of a Navy unwilling to accept change. In fact, a list of leaders supporting employment of aviation as an important weapon included Secretary Daniels; Assistant Secretary Franklin D. Roosevelt; CNO Benson; chief of the Bureau of Construction and Repair Rear Adm. David Taylor; Commander of the Atlantic Fleet Henry Mayo; General Board members, admirals, Charles Badger, Albert Winterhalter, and Frank Fletcher; Commander in Europe Adm. William Sims; Capt. Hutchinson Cone and Capt. Thomas Craven; and various reformers and advocates like Cdr. Henry Mustin and Adm. Bradley Fiske.

The Navy's efforts to mount an aerial assault against the German submarine threat also became the fiery crucible that shaped Marine Corps aviation. Beginning as a tiny assemblage of officers and enlisted personnel in search of a mission and an identity, the flying leathernecks ultimately evolved into a substantial strike force carrying out bombing and resupply missions on the Western Front, thus laying the foundation for all that followed. The aviators' wartime activities earned them a place at the postwar table. Many veterans of the European campaign played significant roles in developing Marine aviation in the interwar period and directing operations in World War II.

Documenting the Northern Bombing Group experience is important for identifying both what was accomplished and what was not. The bombing initiative reflected the Navy's essaying various alternative roles for aeronautics as well as underscoring the heated debate over whether to adopt land- or seaplanes as the basis of naval aviation's future. Studying the group's activities also forms part of the larger discussion of the origins and effectiveness of strategic bombing as a tool of war. Historian Edward Coffman once noted that Gen. William Mitchell, the Navy's postwar nemesis who dreamed of a single air force, should have applauded the Northern Bombing Group (and related programs) for implementing his own theories. These included using "Airpower" to destroy an enemy force or its ability to wage war without directly engaging that force with its opposite number, in this case demolishing the opposing navy's supporting infrastructure without actually engaging that fleet.[3] It is no small coincidence that Robert A. Lovett, the young Navy Reserve officer who emerged as the intellectual inspiration of the Northern Bombing Group, reprised his role in World War II on a much larger scale as Assistant Secretary of War for Air and one of the principal civilian architects of the United States' massive strategic bombing campaign launched against Germany and Japan.

Chapter One
Blazing the Path

The Royal Naval Air Service and the
Beginnings of Strategic Bombing

The winding path that led to the establishment of a major U.S. Navy strategic bombing program in northern France in 1918 began four years earlier in the fertile imagination of Winston Churchill, often considered the "fairy godfather" of the Royal Naval Air Service (RNAS).[1] Under his vigorous and resourceful leadership as First Lord of the Admiralty, the Naval Wing of the Royal Flying Corps (RFC) devoted its efforts to the creation of an air service that was essentially offensive in nature.[2] While the Military Wing focused on reconnaissance, attack became the cornerstone of the Naval Wing.

When war broke out with Germany on August 4, 1914, Churchill was greatly concerned about the bombing threat posed by enemy zeppelins. He had followed the development of these gigantic airships for several years and recommended taking steps to defend England against this weapon as early as October 1913.[3] Zeppelins were difficult to destroy in the air. They could fly higher than the aircraft of the day and were hard to hit using the limited anti-aircraft technology available to gunners on the ground. And if by chance a pilot reached an airship, his options were limited. Grenades would simply bounce off the airship's skin and while rifle fire and machine-gun rounds could puncture the envelope, they would not ignite the hydrogen gas that provided lift unless a spark was created.[4] The best way to counter this new weapon in Churchill's mind was to destroy the bases from which they operated.

As the Germans overran Belgium in mid-August, all the Channel ports were exposed, and the danger of air attacks upon Great Britain became more serious. Zeppelins had already cruised over Antwerp, and it was known that London lay within range of the behemoths housed in giant sheds at Düsseldorf and Cologne.[5] The airship sheds were large structures used for storing the zeppelins, which needed protection from high winds when not in flight, and thus constituted relatively easy targets. The destruction of such structures would disrupt enemy operations. If the shed contained an airship when attacked, so much the better, for it would undoubtedly be destroyed as well. The man selected for this important task was Squadron Leader Spenser Douglas Adair Grey, RNAS.

Spenser Grey began his service in the Royal Navy as a midshipman in 1904, learned to fly at his own expense in 1911, and joined the fleet's nascent air detachment the following year. He became Winston Churchill's favorite pilot and took the First Lord of the Admiralty up in the air on numerous occasions for flying lessons. Grey, who had been promoted to squadron leader upon the official separation of the RNAS and the RFC on July 1, 1914, was called to Admiralty House in London at the end of August to meet with Churchill and the First Sea Lord, Sir John Arbuthnot "Jacky" Fisher, to discuss plans to bomb the zeppelin sheds at Düsseldorf and Cologne.[6] Grey quickly summoned Flight Lieutenant Reginald "Reggie" Marix, RNAS, to join them.

Both airmen expressed their eagerness to "have a crack" at the zeppelin sheds from the RNAS advance base at Antwerp, Belgium, but how, they asked, was it to be done with the flying machines then available? Churchill, aware that two Sopwith Tabloids were going begging at the RFC establishment at Farnborough, volunteered the aircraft. The Tabloids were land versions of the recent Schneider Trophy winner fitted with wheels in place of floats. They were considered very fast for their day and had been sent to Farnborough for trial by the RFC, which turned them down as being unsafe. Would the two men like to try them, asked Churchill. Neither flyer thought that Sopwith Aviation Company, Ltd., of which they were great fans, would turn out an aircraft that was radically wrong. Yes, they answered, they would.

The next day, Grey and Marix traveled to Farnborough where they took the Tabloids up for test flights. The RNAS airmen had no difficulty controlling the biplanes, despite the warnings given beforehand by the RFC pilots who had previously flown the machines. Having ascertained their suitability, Grey had them transferred to the naval air station at Eastchurch where they arrived on September 9.[7] From there they would fly to the base of operations previously established in the besieged city of Antwerp by Squadron Commander Eugene L. Gerrard, one of the Royal Navy's original four pilots.

The first attack on the airship sheds was launched from the Antwerp base September 22, when a two-plane section led by Gerrard took off for Düsseldorf. A second two-plane section led by Grey followed shortly thereafter headed for Cologne.[8] Although the weather was clear in Antwerp when they departed, a thick mist blanketed the ground over Germany making it difficult to locate the targets. Flight Lieutenant Charles H. Collet flying Sopwith tractor no. 906 was the only pilot able to locate his objective. He attacked dropping all three of the 20-lb Hale bombs he carried. The first bomb landed short and the other two failed to explode. Grey, who had been assigned the zeppelin shed at Cologne, used up so much fuel trying to find his target that he had to turn back without dropping any ordnance. None of the other pilots were able to locate their targets and no damage to the enemy was reported.

Bad weather and time spent installing extra fuel tanks in the Tabloids to extend their range forestalled immediate attempts to reschedule the bombings.[9] In the meantime, the Germans had advanced to the outskirts of Antwerp and were threatening to take the city unless the British could send reinforcements. On October 3, 1914, Churchill arrived on site to confer with the Belgian prime minister in order to bolster the city's defenses. He offered to send in the Royal Marine Brigade then at Dunkirk, to be followed by two brigades of the Royal Naval Division, which were still in England training.[10] Churchill remained in Antwerp for several days to see that His Majesty's troops were properly integrated into the Belgian defenses. It soon became clear that it was only a matter of time before the Germans overran the city.

Before heading back to England on the evening of October 6, Churchill discussed plans for a second raid on Düsseldorf and Cologne. In his memoirs, Marix recounts the amusing story of how Spenser Grey convinced the First Lord to allow the two flyers to undertake the raid before the RNAS detachment was ordered away from the threatened city. Marix tells of Grey arguing with the First Lord while the latter was ensconced behind the door of the toilet at British headquarters in the Hotel St. Antoine.[11] Grey won the argument, for the two Tabloids (RNAS Nos. 167 and 168) that he and Marix had flown to Antwerp were left behind when Gerrard's small force was ordered to evacuate the next day.

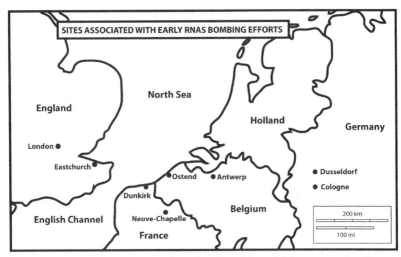

The Royal Naval Air Service began launching bombing raids in 1914 against German airship facilities in Düsseldorf and Cologne.

That night, the Germans began bombarding the city with artillery fire. The airfield that Grey and his fellow airmen had been using stood halfway between the city and the front lines and the shells passed continuously overhead.[12] To avoid damage from splinters if the shed housing their airplanes were hit, the two Tabloids were pulled out and placed in the middle of the airfield. In the morning, the weather was misty and unsuitable for flying, so they spent the time tuning up their machines. The sky had not cleared by 1:00 p.m. and the weather showed no signs of improvement when Grey decided that they must attack before the approaching Germans overran the airfield. Grey lifted off at 1:20 p.m. in No. 167 and headed toward the airship sheds at Cologne. Marix followed ten minutes later in No. 168 with orders to bomb the shed at Düsseldorf.

When Grey arrived over Cologne, he found it blanketed with thick mist. He had been given two different positions for the airship sheds, one to the northwest and one to the south of the town. As he dropped down to 600 feet looking for his targets he came under heavy fire. After ten or twelve minutes of fruitless searching, Grey decided to attack the main railway station. Packed with trains, it made a suitable target of opportunity. He dropped his two bombs on the station and headed back to Antwerp He had an uneventful return flight and landed at 4:15 p.m.[13]

Marix had better luck as he had no trouble locating the airship shed at Düsseldorf. The installation, which had been unsuccessfully attacked by the

RNAS two weeks earlier, was now heavily defended. The intense anti-aircraft and small arms fire put up by the enemy forced Marix into a steep dive as he approached the target. He had dropped to 600 feet when he released the two Hale bombs slung under the Tabloid's wings.[14] One or both bombs must have gone through the shed's roof and exploded inside, for as he pulled up Marix saw enormous sheets of flame pouring out of the huge structure. It collapsed thirty seconds later destroying the Z9 within. The new airship had just completed her acceptance trials and had only been taken inside the shed the day before while awaiting her commissioning ceremony. The raid caused great consternation in Berlin, where such an attack was considered impossible for the British aviators to have conducted. Most historians regard the operation, which was conducted under Grey's leadership, as the first successful strategic bombing mission. Both pilots later received the Distinguished Service Order for meritorious service under fire.

After returning to England at the beginning of 1915, Grey was reassigned to land planes and posted to France for duty with the RNAS Dunkirk Command as C Squadron leader in No. 1 Wing. The wing was situated at the airfield at St. Pol not far from the command's seaplane base at Dunkirk. Combating the submarine threat that had emerged in the autumn of 1914 was one of the primary missions assigned to the command. The ports of Zeebrugee and Ostend in Belgium, now in enemy hands, provided perfect bases from which to launch submarines and destroyer attacks against cross-Channel traffic and ships that passed through the Strait of Dover carrying food and stores to London. To combat this threat, the RNAS began launching bombing raids against submarines based in the Belgian ports, which lay within easy reach of the rapidly expanding airbase at Dunkirk.

Destruction of the zeppelins that had begun to attack the coastal towns of England constituted another high priority mission for the RNAS forces stationed in and around Dunkirk. Aircraft from Dunkirk were sent up to try and intercept the airships as they cruised back to their bases in Flanders. None was encountered until the early hours of May 17, 1915, when LZ 39, which had set out with her two sister ships for a raid along the French and British Channel coasts, was seen off Dunkirk moving slowly eastward.[15] Two RNAS machines were already aloft patrolling when the German airship was spotted. Seven more aircraft, including a Nieuport 11 flown by Spenser Grey, were immediately sent up to try and intercept the zeppelin.

Grey drew abreast of the enemy airship at 9,800 feet, but could not climb any higher; he opened fire on the airship's rear gondola with the Lewis gun mounted on the Nieuport's upper wing.[16] The four machine guns in the gondola returned fire as the airship put her nose up and climbed away. Flight Commander Arthur Wellesley Bigsworth, flying an Avro 504, managed to get above the airship as it moved off in the direction of Ostend. He dropped all four of his Hale bombs on the airship's back and saw smoke coming from the Zeppelin's tail, but was chagrined to see the big airship move off apparently unaffected by the attack. Although the LZ 39 sustained damage in the attack, she was able to make a rough, but safe, landing.

Throughout the spring, summer, and early autumn of 1915, aircraft assigned to the Dunkirk Command provided protective air and reconnaissance patrols over the Channel, while continuing to strike enemy bases and installations at Ostend and Zeebrugee, as well as the airship sheds at Evere and Berchem Ste.

The RFC, which up to this point had focused its attention almost entirely on battlefield reconnaissance, knew little about bombing or its requirements.[17] RFC flyers had experimented with various types of aerial weapons, from grenades to flechettes, but bombing had been left to individual initiatives against targets of opportunity. In mid-February 1915, as it prepared to support the British Expeditionary Force (BEF) in its first offensive, RFC headquarters issued a directive ordering that one flight per squadron specialize in bombing.[18] The only guidance given to the aircrews was to bomb at low altitude. The RFC fielded fewer than ninety front-line aircraft at the time and possessed neither proper bombsights nor releasing gear.

When the BEF mounted its first large-scale attack on Neuve-Chapelle on March 15, 1915, mist and cloud cover impaired air-to-ground cooperation. Although the RFC was able to map the front via aerial photography, there was little coordination between the Army and the air arm. Tactics were decided at the squadron level, the most common practice being aircraft attacking singly at low level. The main objective was to disrupt the enemy's lines of communication. These were the railways used to bring up reinforcements.

The results obtained, according to a report prepared by RFC headquarters, were "in no way commensurate with the efforts made," except for sorties executed by RNAS pilots.[19] The report, issued in July 1915, analyzed 484 aerial attacks carried out between April 1 and June 18, 1915, in which the RFC, RNAS, and the French air service dropped 4,062 bombs. While attacks against naval targets—mostly zeppelins in their sheds—achieved success 25 percent of the time, those against railway stations and junctions—the main targets of the RFC—were successful only 2 percent of the time. RNAS success was attributed to the fact that the efforts of the Dunkirk wing were concentrated almost solely on bomb dropping and that their pilots had been systematically trained for this one purpose. The poor performance of the RFC was vaguely blamed on incorrect bombing methods and the choice of unsuitable objectives.

The evidence presented in the report showed that bomb dropping required specialized tactics and training, more accurate bombsights, improved bomb-release gear, and better aircraft. Instead of implementing these improvements, General Headquarters (GHQ) adopted the attitude that because bombing proved ineffective it was preferable to concentrate on reconnaissance and artillery spotting. Henceforth, stated a directive by the Chief of the General Staff, Army commanders were to restrict bombing attacks by airplanes under their command to specified targets within close reconnaissance area of the Army. Bombing operations to disrupt the enemy's railways would only be carried out under the authority of the GHQ as part of the overall Allied operational plan.[20]

In November 1915, Wing Commander Charles C. Lambe (RN), commander of RNAS forces at the Dover-Dunkirk air bases, sought approval to create two special bombing wings and the airfields needed to support them for use in the wider aerial offensive that he planned to implement in the spring of 1916.[21] Lambe recommended that No. 4 Wing (four squadrons with six pilots in each) be transferred to one of the new airfields, and that the other Wing (No. 5) be formed by detaching four squadrons of six pilots each from No. 1 Wing at Dunkirk. Lambe's recommendation was approved and work on sites at Coudekerque and Petite Synthe began.

No. 5 Wing was formed at Dover under Spenser Grey's command in early March 1916 from personnel serving at Dover and Dunkirk. The unit, which was initially equipped with French bombers—single engine Breguet Vs and twin engine Caudron G.IVs—quickly took up quarters at Coudekerque, France, about four miles southeast of St. Pol.[22] No. 5 Wing carried out its first mission on March 20, 1916, when it joined a combined Allied force of British, French, and Belgian aircraft attacking the German airfield at Houttave and the seaplane base at Zeebrugge.[23] In April, it received a flight of Sopwith 1½ Strutters.

At the end of May, Vice Adm. Reginald H. Bacon, commanding the Dover Patrol, ordered the suspension of the bombing operations being conducted by No. 5 Wing and the other squadrons in the Dunkirk Command.[24] Bacon took this action because he felt that the aerial force under his command was too weak to inflict appreciable damage on German bases, and the light attacks they were conducting alerted the enemy defenses and might bring retaliatory raids against important naval air bases in Dunkirk. With the suspension of bombing operations, every available aircraft fit for fighting was to be placed in readiness for naval operations.[25]

The bombing stand-down lasted until August 1, when the commander of the Royal Flying Corps in France, Major General Hugh M. Trenchard, asked the RNAS to attack the airfield at St. Denis Westrem southwest of Ghent and the ammunition dump at Meirelbeke a few miles beyond. The raid, which involved aircraft from Nos. 4 and 5 Wings escorted by five Sopwith 1½ Strutter fighters, was coordinated with an attack by II Brigade, RFC. After the first raid in support of the Army, Wing Commander Lambe agreed to provide additional bombing support to the RFC, whose resources had been stretched to the limit by horrendous losses incurred as the Battle of the Somme raged some seventy miles south of Dunkirk. In addition to attacking enemy airfields and ammunition dumps in the northern areas of the front, forces under Lambe's command struck zeppelin bases in Belgium and shipyards at Hoboken.[26]

In the last few days of August, Spenser Grey learned that Lt. John H. Towers, a U.S. Navy pilot sent to observe aerial operations in Europe had finally made his way to France. Towers, then serving as assistant naval attaché in London, was one of the pioneering aviators in the U.S. Navy, having qualified as a pilot in August 1911 flying the Navy's first airplane. He arrived at Dunkirk with permission to visit the main air station, but could go no farther. Grey made the short trip from Coudekerque to meet Towers and succeeded in obtaining permission for the American to accompany him back to the aerodrome to observe the operations of No. 5 Wing. Once there, Towers was assigned to Grey's quarters, which were located in a large discarded packing case that had been used for shipping aircraft. These accommodations, which Towers considered to be "quite livable, consisted of a sitting room, two bedrooms, and a bath."[27] While at the field, Towers observed takeoffs and landings for bombing missions conducted on September 2 and 3. He chafed to go aloft himself, but had to wait until the day before he was due to leave before Grey permitted him to go on a raid.

When No. 5 Wing received orders to join the operation against the German airfield at St. Denis Westrem on September 7, 1916, Towers was assigned as an observer in one of the Sopwith 1½ Strutters that would escort the bombers. Takeoffs were normally scheduled before daylight so that the slow, bomb-laden Caudrons would have time to get across enemy lines before daylight exposed them to the anti-aircraft fire. Lingering rain apparently delayed the takeoff, however, for it did not occur until noon when twelve Caudrons and six Sopwiths armed with 65- and 16-lb bombs took off. They were followed by two Sopwiths. Another Strutter joined the flight, but was not listed on the official record. This was Towers' plane, flown by Flight Lieutenant Frank G. Anderae. Grey had no authority to send Towers, a neutral, into combat, and it would have been extremely embarrassing for all concerned if the American officer's plane had been forced to land behind enemy lines.[28] Towers only sighted two enemy machines during the entire mission, but his plane was rocked by anti-aircraft fire. As he recalled in later years, "Several explosions were so close aboard that I could feel the shock and smell the pungent odor of the explosive gases."[29] Thus Towers—thanks to Grey's intervention—became one of the first U.S. Navy airmen to experience combat in World War I.[30]

Grey's No. 5 Wing continued to support the British bombing campaign in Belgium during the course of the Battle of the Somme. As Vice Admiral Reginald Bacon had earlier predicted, the RNAS force was too weak to materially affect the German war effort. The RNAS' primary bomber—the Caudron G.IV—lacked load-carrying capacity (it could carry only two 112-lb bombs per sortie) and the Dunkirk Command possessed too few bombers to be effective.[31] This situation began to change in November when four Short Bombers entered service with 7 Squadron, No. 4 Wing.[32] It had a maximum load of eight 112-lb bombs, but an even larger aircraft was on the way. This was the Handley Page 0/100, which could carry up to fourteen 112-lb bombs.[33]

The Handley Page 0/100, which had a wing span of 100 feet and weighed 14,000 pounds when fully loaded, was a huge aircraft for its day. The first of these arrived in Coudekerque on March 4, 1917, flown by Spenser Grey, who had been promoted to Wing Commander on the last day of 1916.[34] Four more 0/100s arrived in the first week of April and were allocated to 7 Squadron bringing the unit's strength to five Handley Pages and seven Shorts.[35] Initially, these enormous aircraft were used to carry out daylight bombing patrols against enemy destroyer flotillas attacking coastal shipping and bombarding towns along the English coast. The new warplanes drew first blood on April 23 when one of them scored a direct hit on a destroyer with a 65-lb bomb. The enemy struck back three days later when a German floatplane fighter engaged a Handley Page over the Channel.[36] The fighter's machine-gun fire ruptured the bomber's fuel tanks, forcing it to ditch in the open sea. To prevent further losses of this important new asset, Grey decided to restrict the big aircraft to the night bombing role that 7 Squadron had trained for. The first of these raids took place on May 9 when the squadron's Handley Pages bombed Zeebrugge.

Although plans to conduct long-distance raids against targets in Germany had been placed before the Admiralty since the outbreak of the war, the shortage of aircraft and their minimal performance limited the number and type of missions that could be conducted until larger aircraft designed specifically for the bombing role began to arrive at the front. In October 1915, the French advanced the idea of establishing a combined force to raid German industry during the monthly meetings between the British and French air services to discuss questions relating to the supply of aeronautical material. The Admiralty, aware that new and better aircraft were soon to be delivered from Sopwith, Handley Page, and the Short Brothers, responded enthusiastically. The two military organizations soon reached a firm understanding about the need to launch a combined air offensive against Germany proper as soon as enough aircraft became available.[37]

In May 1916, Wing Captain William L. Elders, RNAS, reached France to lay the groundwork for the arrival of the first strategic bombing unit then being formed at the Royal Naval Air Station at Manston. The new unit, known as No. 3 Wing, would operate from an airfield near Luxeuil-les-Bains.[38] This would put it within range of the vital German steelworks located in the Saar Valley. Ongoing zeppelin raids against England played an important part in the decision to hit back at targets in Germany and except for one raid on the town of Freiburg—in reprisal for submarine attacks on two British hospital ships—all of No. 3 Wing's attacks would be made against selected industrial plants or other military targets such as airfields and railroad junctions.[39]

Plans called for No. 3 Wing to be equipped with twenty Sopwith 1½ Strutters and fifteen Short bombers by July 1, 1916.[40] Its ultimate strength was expected to be one hundred aircraft. The Strutters being assigned No. 3 Wing were the single-seat bomber version known in the Admiralty as the Sopwith Type 9700. The Type 9700 was converted into a bomber by replacing the standard two-seat Strutter's rear cockpit with an internal bomb bay capable of carrying four 65-lb bombs.[41] Although the Shorts were designed to carry a larger ordnance load—either four 230-lb bombs or eight 112-lb bombs—they proved to be unsuitable in service and were soon replaced by the Breguet V.

No. 3 Wing began operations on July 30, 1916, when a small combined French and British force took off to attack Mulheim a few miles northeast of Cologne.[42] Bombing on a large scale was curtailed until October, however, due to a shortage of aircraft when the Battle of the Somme began to strain the resources of the RFC. The War Office made an urgent appeal to the Admiralty to hand over as many aircraft as the RNAS could spare. Trenchard, then in command of the RFC, asked for and was promised sixty 1½ Strutter fighters as soon as they were available. Fighter production became a priority at the Sopwith works, curtailing output of bombers that had been destined for the Luxeuil aerodrome. As a stopgap measure for the 1½ Strutters that the wing should have had, the Admiralty provided a number of French Breguet bombers. These pusher biplanes proved markedly inferior to the Strutter and were not popular with naval pilots who considered them unsuited for daylight operations. Unpopular or not, the Breguets arrived in time for the first large-scale attack made against the Mauser arms works at Oberndorf on October 12, 1916.[43]

Additional army demands for aircraft and personnel further hampered RNAS operations. At the end of January 1917, Captain William Elder, RN, was ordered to transfer nine of his best pilots to Dunkirk where they were needed to assist the RFC. No. 3 Wing lost nineteen more pilots, six 1½ Strutter fighters, and one hundred ratings on March 7 when they too were posted to Dunkirk, causing another major reorganization. Three weeks later, on March 25, Elder received a telegram from the Admiralty instructing him to close down the wing.[44] The eviscerated unit had only been able to mount thirteen raids between October 1916 and April 1917.[45] The average number of aircraft participating in each mission was fifteen and only 2,500 pounds of bombs were dropped per raid.

The demise of No. 3 Wing and suspension of strategic bombing against Germany resulted from the ongoing feud between the War Office and the Admiralty over scarce aviation resources. It was initiated on February 4, 1915, by the Director-General of Military Aeronautics, Brigadier General Sir David Henderson, when he complained to the Joint War Air Committee about the unfair methods being used by the Admiralty to secure aero engines. At the beginning of the war, the Admiralty had been given priority in the supply of high-powered airplane engines because of the extra power needed by seaplanes. Now, the RFC, which had an urgent need for such engines, was having great difficulty obtaining them because the Admiralty was still placing orders with firms that the two air services had mutually agreed to allot to the RFC. Henderson asked the committee to decide the best allocation of resources based on the duties defined for each service and their relevant importance to the war effort. What Henderson wanted, according to the analysis presented by Neville Jones in *The Origins of Strategic Bombing*, was to obtain a ruling giving the RFC sole responsibility for land operations so that supplies of aircraft and aero engines being assembled for strategic bombing would be reallocated to the Royal Flying Corps. A second memorandum sent by the War Office to the committee on March 2, 1916, accusing the RNAS of interfering with the development of the military wing set off a firestorm within the Admiralty.[46]

Trenchard entered the fray in June 1916. Writing to the Air Board on June 9, he proclaimed the critical importance of providing observation for accurate artillery fire that was essential to the success of ground operations. If resources were not sufficient to provide forces for both tactical support and strategic bombing, then the latter must be discarded in favor of the former.[47] Thus, No. 3 Wing was sacrificed to enable Trenchard to throw greater numbers of aircraft into the coming air battles on the Western Front. Even after the Battle of the Somme ended, Trenchard pressed for more material from the RNAS. On December 11, 1916, he appeared in person before the Air Board asking that four fully equipped naval squadrons and 150 aero engines being allocated to the Admiralty be transferred to the RFC. The Air Board put his "request" to the Admiralty who agreed to assist.[48]

The dispute between the War Office and the Admiralty over the latter's use of resources to pursue its bombing policy reached a head in October 1916, when Colonel Joseph Èdouard Barès of the French air service and an ardent supporter of strategic assaults came to London seeking British support for an extension of air attacks against Germany. Barès appealed for greater cooperation between the two countries, arguing that the Allies should embark on a campaign of reprisal raids on towns to be carried out immediately after a zeppelin raid or submarine attack. He believed "the end of the war would be brought about by the effective bombing of open towns" and that a large bomber force would be needed to achieve this.[49]

Although the Admiralty did not agree with Barès' policy for assaulting population centers, it enthusiastically endorsed his proposal for extending the Anglo-French bombing effort and asked the Air Board for authority to provide "an effective force of at least two hundred bombers in France."[50] Henderson's response was predictable; he rejected the proposal outright. Taking another shot at the Royal Navy, he declared that bombing "is an unimportant duty compared with fighting or reconnaissance, nor does it seem to fall in any way within the sphere of naval duties." Trenchard agreed.[51] He appealed directly to Lieutenant General Douglas Haig, commander in chief of the BEF, "to throw the weight of his authority into the attack on the naval plan."[52] Haig complied, and via a strong letter of protest dispatched from GHQ France attacked the Admiralty's plans on the grounds that they were based on mistaken air policy. According to the army chief, there was only one way to bring about the successful end to the war and that was to obtain "a decisive victory over the enemy's forces in the field."[53]

The first debate over the RNAS role in strategic bombing was resolved in favor of the army after the *de facto* acceptance of Haig's attitude toward the division of responsibilities and the importance of observation missions. After conducting its last attack against the city of Freiburg—in retaliation for two U-boat attacks on unarmed hospital ships on April 14, 1917—No. 3 Wing disbanded and its pilots and aircraft redeployed to assist the RFC elsewhere on the Western Front. By then the United States and the Allies had begun planning to capitalize on America's huge industrial base in an effort to make up for shortfalls of personnel and production of airplanes and aero engines.

Though RNAS efforts to maintain a targeted bombing offensive had fallen short due to institutional and operational disagreements with the British Army and the Royal Flying Corps, the initiatives it carried out in 1914–17 laid the groundwork for the U.S. Navy's subsequent efforts in this field. In fact, the Northern Bombing Group program that emerged in the spring of 1918 specifically picked up the strategic torch reluctantly set down by the RNAS. It would be on these foundations that the Navy would erect its own program, its largest offensive initiative of the war.

Chapter Two
Crushing the Hornets' Nest

The specific series of events that culminated with the United States Navy's improbable creation of a strategic bombing force in the spring and summer of 1918 commenced in the winter of 1917 in the cold waters of the North Atlantic when Germany made the fateful decision to unleash unrestricted submarine warfare against the Allies and their neutral suppliers. All American vessels were now at risk. Sinkings by U-boats jumped to 540,000 tons in February, nearly 600,000 in March, and 874,000 in April. A crisis was at hand.[1]

Responding to Germany's unchecked U-boat depredations, the United States declared war in early April 1917, but without plans, experience, personnel, equipment, or facilities to battle the underwater scourge. President Woodrow Wilson and several Navy leaders, however, desired an active campaign against enemy forces, taking the war to the hornets in their nest, something they believed the British fleet had thus far declined to do. The Navy Department quickly set about gathering information and formulating plans to defeat the submarine.

The severity of the submarine threat colored all discussions concerning America's future naval strategy. At the commencement of unrestricted U-boat operations, Germany possessed approximately one hundred undersea vessels, including forty-eight based at North Sea ports and two dozen in Belgium, with perhaps three dozen on patrol at any given time. According to one analyst, Germany's declaration of unrestricted submarine warfare "shocked" Washington. At first, President Wilson, Secretary of the Navy

Josephus Daniels, and Chief of Naval Operations William S. Benson hoped to remain above the conflict, with the President contemplating various measures short of war. Professor Frank Freidel, Franklin Roosevelt's biographer, asserted that both Wilson and Daniels wished to avoid any action that might be interpreted by Germany (and historians) as aggressive, and that a cautious, deliberate Benson carried out their wishes. Daniels himself later wrote, "From the beginning of the war in Europe, I had resisted every influence that was at work to carry the United States into war." At the same time, the General Board of the Navy defined the fleet's role as protecting the lives of Americans on the high seas and recommended warships be deployed against the submarines. Comparatively ill-informed concerning the growing threat, Washington remained focused on defending the Americas and not provoking Germany. This situation persisted for nearly seven weeks.[2]

Nonetheless, the President urged that everything possible be done to protect United States shipping. On March 6, Wilson visited Daniels in his office to discuss the issue of arming merchant vessels, the dangers of submarines in American waters, and bringing the fleet northward from the Caribbean. They continued their conversation at the White House on March 8.[3] Orders to arm merchant ships were issued a few days later. As late as March 19, however, Daniels claimed Wilson "still hoped to avoid [war] and wished no effort [be] spared to protect shipping, putting efficiency above prudence." The Secretary also promised to call the General Board into session "to consider every method to protect our shipping, it being paramount, and send him their report." At the same meeting, Rear Adm. James Oliver, director of Naval Intelligence, received instructions to "learn what [the] English were doing to stop or lessen submarine warfare."[4]

As losses at sea mounted, however, descent into war became inevitable. On Sunday, March 18, Daniels recorded, "Awful night. Three American ships reported to be torpedoed by German submarines."[5] Two days later, Wilson and his cabinet decided to ask Congress for a declaration of war. That same day Assistant Secretary of the Navy Franklin D. Roosevelt and Rear Admiral James Oliver contacted Captain Guy Gaunt, British naval attaché in Washington, requesting his government's views on the issue and looking for a plan of cooperation. The following day, the Navy Department approved a plan to mobilize the fleet in case of war.[6]

During these final weeks of peace, the pace of activity at Navy headquarters accelerated. Rear Adm. Albert Gleaves requested more men and aircraft be assigned to the Destroyer Force for protection of the fleet. Battleships were ordered north to Chesapeake Bay. Orders were placed for construction of additional destroyers. President Wilson authorized increasing Navy enlistments by 57,000, including 200 new aviation ratings. A decision was reached to recruit women, for certain clerical duties, as Yeoman (F), the first time in Navy history that females had been authorized to serve. In Connecticut, a group of college aviators known as the First Yale Unit traveled to New London and enlisted en masse. The flight school at Pensacola went to double-shift instruction. Additional test flights to detect submarines were performed.

On March 25, Roosevelt and Frank Polk, Counselor for the Department of State, met with Captain Gaunt and invited the British to send a delegation to the United States to brief the Americans on anti-submarine measures and coastal patrols. In fact, as early as March 24, an internal Admiralty memorandum listed four types of naval assistance the United States could provide. Two days later, the War Cabinet discussed the need to dispatch an official to Washington to facilitate Anglo-American cooperation.[7]

About the same time, President Wilson raised the issue of sending "Naval officers of high rank to England and cooperate to protect our merchant ships and ask them to send officers here." Daniels first selected Capt. Henry Wilson who demurred, preferring sea duty. Daniels then chose Rear Adm. William S. Sims, newly appointed head of the Naval War College in Newport, Rhode Island. In his postwar memoir, *The Victory at Sea*, Sims described his mission as getting in touch with the British Admiralty, studying current conditions, and learning how the United States "could best and most quickly cooperate in the naval war."

During Daniels' discussions with Sims prior to his departure the Secretary revealed reservations concerning the admiral's selection, observing, "You have been selected for this mission not because of your Guildhall speech, but in spite of it."[8] Here he referred to an impolitic address Sims had given in England in 1910 for which he received an official rebuke from President William Howard Taft. Daniels also discussed a confidential assessment from Ambassador Walter Page that the submarine threat was much more serious than the British were admitting. Speaking directly for the President, Daniels relayed Wilson's belief that the Royal Navy had not mounted a vigorous offensive against the U-boats and every effort should be made to prevent submarines from transiting from their bases to the open waters of the Atlantic. In addition, the President, Daniels, and the General Board all favored convoying merchant ships. Outfitted in civilian garb, with Daniels' instructions fresh in his mind, Sims and his aide, Lt. Cdr. John V. Babcock, crossed the Atlantic on board the liner *New York*, reaching England April 9, three days after Congress declared war on Germany.[9]

From his new perch in London, Sims bombarded the Department with cables highlighting the crisis unfolding overseas. He claimed the magnitude of the submarine threat had been underestimated and urged all possible assistance, especially immediate dispatch of any and all available destroyers. On April 14, Sims sent a cable to Washington that Daniels described as, "So confidential he [Sims] sent it in the State Department secret code" fearing that it might help the enemy morale if known.[10] "It will be delivered tomorrow." Admiral Sims informed Daniels, "The submarine issue is very much more serious than the people realize in America. The recent success of operations and the rapidity of construction constitute the real crisis of the war. . . . The reports in our press are greatly in error. . . . Supplies and communications of forces on all fronts, including the Russians, are threatened and control of the sea actually imperiled." Additional communications from Sims reinforced this assessment, calling the situation presented by the submarine as "not only serious but critical. . . . Briefly stated, I consider that at the present moment we are losing the war." Dispatches from Ambassador Page buttressed these assertions.[11]

These claims did not immediately sway opinion within the Navy Department. Wilson, Daniels, and Benson remained suspicious of both Sims' and Page's objectivity. In a cabinet meeting held earlier on March 28, Wilson had declared, "Page meddles in things outside his domain. I do not mind this if he gave us his own opinions, but he is giving him [*sic*] English opinion." Benson even informed Sims "there was a feeling that he was being unduly influenced by the British." At one point, President Wilson said of his commander in Europe, "Admiral Sims should be wearing a British uniform." According to historian William Still, "Secretary Daniels and Admiral Benson deplored his [Sims] Anglophilism." Following the war when Daniels and Sims became embroiled in a bitter controversy regarding the state of the Navy's preparedness in 1917, the Secretary charged that Sims suffered from "the peculiar malady which affects a certain type of American who go abroad and become in many ways un-American." This was a remarkable charge to level in a period of hyper-patriotism.[12]

Objective or not, Sims' warnings could not be dismissed. A few days later on April 21, Daniels answered a telephone call from Secretary of State Robert Lansing announcing receipt of a telegram from London so confidential that he could not talk about it over the phone. Sims reported that thirteen ships had been sunk in the past 24 hours, and losses in the first 18 days of April aggregated 408,000 tons. He called the situation critical and growing worse, and again appealed for destroyers. Daniels immediately called Admiral Benson. Despite their reservations about Sims' objectivity, the magnitude of the threat was becoming obvious. The question of dispatching destroyers to Europe had first been raised on April 12. President Wilson approved the concept April 17 and six destroyers set out for Europe on April 24.[13]

Sims' communications also highlighted the dangers presented by the deadly U-boat nest in Belgium. Following the war he wrote, "When the Germans captured the city of Bruges they transformed it into a headquarters for submarines; here many of the U-boats were assembled, and here facilities were provided for docking, preparing, and supplying them. Bruges was thus one of the main headquarters for the destructive campaign which was waged against British commerce."[14] It was this complex of inland docks and naval facilities connected by canals to the coastal ports of Ostend and Zeebrugge that the Navy would eventually confront, first by establishing NAS Dunkirk to patrol water access to these sites and later by creating the Northern Bombing Group to attack them directly.

As shipping losses in European waters skyrocketed, prevailing attitudes among the Allies held that failure to control enemy marauders might cost them the war. According to historian David Trask, no problem created more anxiety in Washington.[15] Throughout May and June, Sims labored ceaselessly to bring Washington over to his view of the U-boat situation and advocated actions the United States should take to counter the threat. By late June, the Navy had dispatched twenty-eight destroyers to Ireland with many more to follow, largely stripping the fleet of its protective screen.[16]

Eventual official support for an aggressive bombing program, something not even imagined in April 1917, resulted from a convergence of factors relating to the Royal Navy's apparent inability to stem the flow of U-boats. In the first six months of the war, President Wilson, Secretary Daniels, CNO Benson, and others (naively) asserted the British lacked boldness in confronting the German fleet or combating the submarine menace. As early as April 16, Benson questioned why the Royal Navy did not blockade the German coast. Strongly suspicious of Great Britain, he doubted its fleet was up to the task of performing its duties properly. Later that month, Sims in London received word the Department was showing interest in close mining operations. A July 3 communication from Daniels to Sims outlining American naval policy identified offensive operations as one of six fundamental objectives.[17]

Throughout this period, President Wilson complained repeatedly of a lack of news regarding anti-submarine measures, growing increasingly testy over the Royal Navy's seeming failure to curb the U-boat. In early July, he communicated these feelings directly to Sims, calling for offensive tactics against the submarine threat and revealing his irritation with British actions thus far, as well as his surprise at the Royal Navy's failure to use its great superiority in such a campaign. Rather, America's new ally seemed helpless to the point of panic. "Every plan we suggest," he carped, "they reject for some reason of prudence." Instead, the President urged new, effective departures in the undersea war. Possible actions included attacks on the German fleet and its ports, including Zeebrugge, convoys, and closure of the North Sea with nets and mines.[18]

Such critical attitudes, hardly a secret in Britain, stirred considerable resentment in London. In June 1917, inventor Arthur Pollen visited the United States, hoping to sell his Argo fire control system to the U.S. Navy. Author John Buchan, then serving as Director of Propaganda in the British Foreign Office, approached Pollen and asked him to publicize the Royal Navy's accomplishments while in America. During his visit, Pollen worked to counter the notion that the British had failed to defeat the German fleet, or resolve the submarine issue, thus leaving American ships open to attack. Pollen strongly supported the dispatch of Rear Adm. Henry T. Mayo, commander of the Atlantic Fleet, to Europe and energetic offensive action. In a similar vein, on July 5, 1917, newspaper baron Alfred Harmsworth, Lord Northcliffe, then heading the British War Mission to the United States, claimed charges of supposed Royal Navy inactivity greatly hindered Anglo-American relations.[19]

When Wilson visited the Atlantic Fleet at Hampton Roads on August 11, his frustration over the submarine question erupted into public view. With the sun shining and the full panoply of the American battle fleet spread before him, Wilson repeated his charges that the Royal Navy had showed little initiative or imagination in combating the underwater threat. He now called for "something unusual . . . something that was never done before." The President exhorted the leaders of the fleet, "Do not stop to think about what is prudent for a moment. Do the thing that is audacious to the utmost point of risk and daring." Wilson informed three hundred officers gathered on board *Pennsylvania*, Admiral Mayo's flagship, "We are hunting hornets all over the farm and letting the nest alone. None of us knows how to go to the nest and crush it, and yet I despair of hunting for hornets all over the sea when I know where the nest is and I know that the nest is breeding hornets as fast as I can find them." Telling the assembled officers, "I know the stuff you are made of," he pronounced himself willing "to sacrifice half the navy Great Britain and we together have to crush that nest, because if we crush it the war is won." One wonders what his audience thought of that prospect.[20]

Just five days later, Wilson returned to the topic, lecturing Daniels, Benson, Mayo, and Capt. Richard H. Jackson on the inactivity of the Allied fleets and the necessity of finding and destroying the hornets' nest through offensive means. The war against the submarine could not be won through hunting around the great ocean, he argued. Instead, Wilson desired plans for a naval campaign that would employ American ingenuity, boldness, and dash, qualities he believed the Royal Navy sorely lacked.[21]

So exercised did the President become on this issue (and so taken with his own insect metaphor) that he informed trusted adviser Col. Edward M. House that he wished to send a mission to Europe "to find a way to break up the hornets' nest and not kill hornets over a forty acre lot." House recorded Wilson again saying he was willing to lose half the Navy to do it. In September, Wilson called for a plan, any plan, to end the submarine menace, whether the British agreed or not. He believed his wartime partners had fallen into a rut. A few in the Navy Department leveled the same criticism against their own leaders. Especially vocal was Assistant Secretary Franklin Roosevelt. He advocated reducing the influence of the "conservative" General Board and placing younger, more aggressive officers in charge. Roosevelt's targets included both Daniels and Benson.[22]

Based on these frequently voiced concerns, Wilson decided to dispatch Admiral Mayo to London to stress the United States' desires in person. When the admiral seemed somewhat cautious about the likely results of the proposed mission, Secretary Daniels mused, "Is Mayo hopeful enough?" Nonetheless, the President strongly wished the mission to Europe to proceed for the purpose of improving relations with the Allies and making the United States' concerns known, as well as advancing claims that the Americans be regarded as "the senior partner in a successful naval campaign." As Daniels informed Assistant Secretary of State William Phillips, the Americans were "tired of playing second fiddle to the British by meeting all their demands."[23]

Mayo departed for Britain at the end of the summer, reaching Liverpool and then London in late August. During the crossing, he organized his thoughts, taking guidance from a memo written by Benson (and reviewed by Daniels and Wilson) and preparing a list of pointed questions for his hosts. What had been accomplished thus far? What initiatives were under way? What programs were planned for the future? Specific concerns included anti-submarine measures, the merchant shipping situation, troop transports, political questions, and the use of aircraft. According to his instructions, Mayo should push for a more aggressive policy and negotiate "from the position of a senior partner."[24]

Mayo lost no time informing the Royal Navy of Wilson's desire for boldness. "You cannot make omelets without breaking eggs," the President had said. "War is made up of taking risks." The naval conference met September 4 and 5 and addressed a wide range of subjects, including attacks on submarine bases and constructing mine barrages in the North Sea and Straits of Otranto, but resulted in few specific decisions. As Mayo reported later, "While the conference was very useful for all concerned, it is extremely difficult to reach any conclusions, other than those of a very general nature." Following the conclave, Mayo toured Royal Navy centers at Portsmouth and Southampton, conferred with officials in Paris, visited the front lines near Amiens, boarded a British warship to observe the bombardment of Ostend, and inspected American forces at Brest and Queenstown.[25]

Despite these tentative steps, Mayo's visit did not end American criticism of British efforts or soothe anxieties over Allied tactics for defeating the U-boat. Mayo, himself, disparaged the Admiralty's failure to provide coherent data concerning their anti-submarine operations and seeming lack of an overall plan of action. Instead, it appeared the naval war had been "carried out from day to day and not according to the effective co-ordination and co-operation of efforts against the enemy." Nonetheless, Mayo believed progress had been made and it was up to the United States to "make the earliest possible decision as to what forms and extent the assistance to be given shall take" and then implement such assistance as quickly as possible. For his part, Wilson remained skeptical, dismissing a British proposal to sink block ships in a North Sea channel, preferring "some real offensive." According to Daniels, who attended an October 19 meeting between Mayo and Wilson, the President returned to his favorite metaphor, noting that even before the United States entered the war he believed the submarine scourge would be ended only by "shutting up the hornets in their nest."

The continued unsettled nature of Anglo-American naval relations, and the larger issues besetting overall Allied-American cooperation in the areas of military policy, finance, shipping, trade, food, and wartime production, led to the dispatch to Europe of Colonel House and a substantial delegation of officials in late October 1917 to attend a large planned conference of the war-making powers. Maj. Gen. Tasker Bliss, Army Chief of Staff, and Admiral Benson represented the United States military. Crossing the dangerous Atlantic in the cruiser USS *Huntington*, the party reached Plymouth, England, November 7.

During the days and weeks that followed, Benson met often with British naval officers and Admiralty officials and forcefully presented Washington's views. He stated clearly that the United States wished to help in any way possible, but also needed to know what future plans were being formulated, wished to see a definite course of action established, and urged that a specific role be assigned to each member of the fighting coalition. Benson also indicated the United States wished to help develop those plans, directives that would then be approved by all participants before being implemented. According to Mary Klachko, the admiral's biographer, Benson "communicated some of the Navy Department's sentiment that the Royal Navy had not been sufficiently inventive and aggressive." Additional meetings offered further opportunities for the CNO to report "American dissatisfaction with the policies of the Admiralty."[26]

Benson used much of his time to urge acceptance of certain initiatives designed to challenge the German submarine more aggressively, including closing the Straits of Dover, constructing a North Sea mine barrage, and mounting direct attacks against U-boat bases (striking the hornets' nest). Moving to France in late November, Benson again attended a seemingly endless series of meetings, held discussions with Allied and American naval officials, and made numerous trips, including a visit to General Philippe Pétain's headquarters at Compiègne. During his time in France, Benson consulted with Admiral Ferdinand-Jean-Jacques de Bon, French chief of the naval staff, and carried out a wide-ranging tour of American naval facilities. He also met with U.S. Navy officials.

Benson's visit to Europe produced several tangible results. The Navy ordered a squadron of battleships dispatched to join the Grand Fleet. The North Sea mining barrage idea moved from vague proposal to the feasibility study phase. A Planning Section was authorized for Sims' headquarters in London—a group that later played a significant role in discussions leading to creation of the Northern Bombing Group. Also important was what Benson did not achieve. There would be no close-in blockade of the German coast. Neither would there be a massive naval assault on the enemy fleet and its shore facilities. The Royal Navy would continue its policy of watching, waiting, and containing. Despite bluster from both Wilson and some of his naval officers, the American battle fleet was in no position to change British policy or initiate bold actions on its own. It was this vast gulf between American priorities and capabilities that aviation soon found itself occupying.

Chapter Three
Naval Aviation
Enters the Arena

April–December 1917

D espite diplomacy, negotiation, and bluster, it became obvious by the end of 1917 that the main elements of the United States battle fleet would play little or no role crushing the U-boat, either at sea or in its lair. There would be no raids, assaults, bombardments, or close-in blockades. Instead, a series of defensive measures was implemented. A squadron of American battleships would reinforce the Royal Navy. Destroyers, cruisers, and obsolete battleships, along with a gaggle of converted yachts and diminutive wooden subchasers would escort convoys and patrol for U-boats. A North Sea mine barrage was in the offing. For the time being, the hornets would remain undisturbed in their nest.

Or would they? There remained one military asset, untried to be sure, that might be deployed to fulfill President Wilson's mandate for offensive action— naval aviation. Still in its infancy, aviation thus far had functioned as a minor, experimental adjunct to the fleet, useful perhaps for scouting, spotting, and patrolling, but not a significant offensive weapon. That was about to change. In less than a year, aviation activities moved to encompass anti-submarine patrolling, offensive hunting missions, and a major strategic bombing initiative. The effort eventually included more than forty thousand personnel, two thousand aircraft, and approximately four dozen bases and training facilities stretching from England, Ireland, France, and Italy, to Halifax, Canada; Key West, Florida; Coco Solo, Canal Zone; and San Diego, California.

Of these aviation activities the most innovative were plans to implement large-scale bombing of enemy bases, facilities, and infrastructure. In fact, even at a distance of nearly a century, the Navy's efforts to implement a significant aerial bombardment program during World War I remain impressive in their size and scope. Had all of the strategic bombing plans been fully realized, the Navy would have deployed more than 1,500 bombers and as many as 15,000 men—a massive commitment by any standard. They also would have taken aviation far from its limited prewar vision of directly aiding the fleet, or its wartime assignment of stalking the U-boat. These plans meshed perfectly with the Navy Department's (and President Wilson's) oft-expressed view that naval forces should undertake more aggressive action against the enemy.

In a very real sense, the Navy's foray into the infant field of long-range bombing was an aberration, a detour from the path followed ever since Eugene Ely first took off from the cruiser *Birmingham* in 1910. The thinking had always been to marry aircraft with the fleet. Early naval aviators operated a heterogeneous mixture of flying boats and pontoon seaplanes, but not land-based machines. Before the outbreak of war, aviators and officials, including Capt. Mark Bristol, Cdr. Henry Mustin, and Lt. Kenneth Whiting, had proposed construction of one or more seaplane carriers. During the same period, the Navy installed catapults on three warships and also conducted scouting, observation, and submarine spotting exercises.

A report prepared March 12, 1917, by the Special Army and Navy Board outlined the roles to be played by the respective aviation forces, with the sea service's responsibilities defined as operating in conjunction with fleets, from shore bases for oversea scouting, and under the commandants of Naval Districts, and advanced bases. More specifically, "The mission of Navy aircraft operating from shore stations is to scout for and report movements of enemy forces at sea; to attack enemy forces at sea; and to assist the Army when operations of the enemy are in the immediate vicinity of the coast." This approach exactly mirrored prevailing Navy Department attitudes at the commencement of hostilities. The Navy believed its primary responsibility consisted of defending the United States and its home waters against the possibility of attack or invasion, hence the need to concentrate the battle fleet and its auxiliaries, including the defensive destroyer screen there. Benson himself later wrote, "My first thought . . . was to see that our coasts and our own vessels and interests were safeguarded."[1]

A memorandum prepared for Secretary Daniels in late June 1917 reinforced existing aviation doctrine. Most likely written by Lt. Cdr. John Towers over Capt. Noble Erwin's signature, the document noted that the Department had consistently pursued the objective of developing aircraft for use with naval forces, particularly shipboard scouting and "controlling the fire of our own guns and opposing enemy aircraft being used for a similar purpose."[2] Other possible activities included overwater patrols and cooperation with Naval District surface vessels in combating enemy surface or submarine craft.

Concrete efforts in the United States both before and after the April 1917 declaration of war embraced exactly these objectives. The Navy's only flight school at Pensacola trained seaplane pilots exclusively, a mission it pursued throughout the war. Other prewar activities included dispatching officers to sites along the eastern and Gulf coasts to identify locations for patrol bases to be used for scouting against possible submarine incursions. Nowhere did the Towers/Irwin paper or other early plans mention or even hint at sustained bombing of enemy naval facilities, attacks on the opposing fleet, or strategic aerial assaults designed to break civilians' morale.

Following the outbreak of war, the Navy concentrated on activities designed to expand and strengthen its aviation arm, but not create a strategic bombing capacity. In early July, the *Air Service Journal* quoted Daniels describing the rationale behind a request for a new appropriation of 45 million dollars for naval aviation, again stressing the now traditional missions envisioned. "The value of aircraft," he said, "has been abundantly demonstrated in the present war and the Navy is making efforts to build up an air force of sufficient size to operate as scouts from naval vessels, to patrol the waters off the extensive coasts of the United States and our insular possessions, and to co-operate with naval forces abroad in anti-submarine warfare." The requested funds would go toward maintaining and expanding existing schools and stations, establishing new stations and training schools, and purchasing necessary aircraft—seaplanes, dirigibles, and kite balloons.[3]

At first, several small, ad hoc training programs instructed groups of college volunteers from Yale, Harvard, Princeton, and other universities. Two dozen student aviators trained in Canada. New instructional centers opened at Bay Shore, New York, and Key West and Miami, Florida. The inadequate Pensacola facility initiated a major expansion program. Significantly, all students learned to pilot seaplanes and small flying boats. Construction of large coastal patrol stations moved ahead at places like Chatham, Massachusetts; Montauk and Rockaway, New York; Cape May, New Jersey; and Hampton Roads, Virginia. The Navy also placed substantial orders for hundreds of aircraft, such as Curtiss N-9 and R-6 pontoon machines. Work commenced on the Naval Aircraft Factory in Philadelphia to speed production. All of these efforts were designed to train pilots, secure aircraft, and construct coastal patrol bases to defend American shores.[4]

Such limited objectives, however, did not long survive firsthand assessment of developments in Europe, the continuing evolution of the enemy threat, or the technological revolution under way. Over the battlefield, simple, sporadic reconnaissance sorties quickly gave way to continuous, sophisticated aerial photography and artillery spotting. High-performance scouts engaged in fierce air-to-air contests. Ground attack machines strafed and bombed trenches, artillery emplacements, supply convoys, and railroad lines. Long-range bombers and zeppelins attacked cities and industrial sites. Aircraft operated over water as well, patrolling for submarines, spotting for the fleet, and on a few occasions, launching successful torpedo attacks.

To help remedy the Department's profound ignorance of wartime aviation needs and conditions, and make a morale-boosting gesture to its new allies, the Navy dispatched the tiny First Aeronautic Detachment to France in late May 1917.[5] Consisting of seven officers, of whom four were aviators, and 122 new bluejackets, the Detachment constituted the first organized, albeit untrained, American aviation force to land in Europe. Formed at Pensacola in early May 1917, the unit soon traveled to Baltimore and New York, where the men boarded colliers *Neptune* and *Jupiter* for their journey to France. Lieutenant Kenneth Whiting, an aggressive and thoughtful junior officer, led the group. Based on his observations he authored a series of recommendations that set the course naval aviation would follow in France for the remainder of the war. Out of these would evolve a string of more than a dozen seaplane, dirigible, and training facilities along the west coast of France stretching from northern Brittany nearly to the Spanish border. Whiting proposed further, that the Navy establish an air station at the exposed, front-line outpost of Dunkirk, from which aircraft could attack enemy submarines in the North Sea and the English Channel. He also recommended launching a bombing campaign against U-boat facilities in nearby Belgium.[6]

A pioneering submariner turned aviator, Whiting was born in Stockbridge, Massachusetts, in 1881, but relocated with his family to Larchmont, New York, shortly thereafter. He entered the U.S. Naval Academy in 1901 and graduated in 1905. During his years at Annapolis he excelled as an athlete—football, hockey, track, boxing, swimming, sailing—and was president of the Athletic Association. Studying, however, exerted no such appeal and he graduated near the bottom of his class. According to Rear Adm. George van Deurs, Whiting possessed a friendly smile, modest manner, and innate ability. "Anything routine bored him. He lived to enjoy the exciting and the unusual and to achieve the impossible."[7]

Early sea duty included time on board the armored cruiser *West Virginia* before being assigned to the gunboat *Concord* and steamer *Supply*, both of the Asiatic Station. In late 1908, the adventurous ensign transferred to submarine duty, first taking command of *Shark* and then fitting out and assuming command of *Porpoise*. In April 1909, he conducted a dangerous experiment during which he ordered the boat submerged to a depth of 20 feet, then crawled into its single 18-inch torpedo tube, and after flooding the space, managed to wriggle out and make his way back to the surface, all in about 75 seconds. Always a daredevil, Whiting also survived a court-martial in this period purportedly for drunkenness.[8] Neither escapade appears to have harmed Whiting's career. He returned to the United States the following year to take command of the submarine *Tarpon* and later the *G-1* boat.[9]

As early as 1910, Whiting and fellow submariner Theodore "Spuds" Ellyson applied for flight instruction with pioneer aviator and aircraft manufacturer Glenn Curtiss. Ellyson was accepted and ultimately earned the designation of Naval Aviator No. 1. The Navy turned down Whiting's request, however, considering him "too slow," and he continued submarine duties until June 29, 1914, when he was ordered to the Wright school in Dayton, Ohio. He passed successfully through the course, the last man personally instructed by Orville Wright, and was designated a naval aviator in September 1914. As with so many other endeavors, "flying was just something else he did well, like swimming or sailing."[10]

Whiting then moved on to the Navy's pioneer air station at Pensacola, Florida, serving for some time as officer in charge and becoming a charter member of the "Old Tribe" associated with that facility. The group earned a well-deserved reputation for flying bravely and playing hard. "When flying was over," noted van Deurs, "drinking was pleasanter than worrying about routine problems that could not be solved with the means available."[11] In November 1916, Whiting transferred to the cruiser *Washington* (soon renamed *Seattle*) as commander of its seaplane detachment. While at sea, he proposed acquisition of a seaplane carrier, to be developed from existing railway ferries, a proposal endorsed by Admiral Gleaves and Admiral Mayo.[12]

The outbreak of war saw Whiting separated from duty on board *Seattle* and ordered to Washington, D.C., to receive instructions regarding the newly created First Aeronautic Detachment. Guided by very vague and open-ended orders, he immediately entered into a busy series of meetings with French military officials that quickly spawned several inspection tours along the coast to identify possible sites for American anti-submarine patrol stations. Whiting also arranged for his men to receive instruction as pilots and observers at various Army and Navy schools. Many would go on to play significant roles providing a cadre of trained personnel for the Northern Bombing Group. The activities Whiting set in motion, especially choosing Dunkirk as an important seat of naval aviation activity, led directly to the Navy's later strategic bombing initiative.

Whiting's tours and recommendations also initiated a pattern that persisted throughout the American wartime experience, wherein Allied needs and objectives shaped Navy decisions. In essence, Whiting went where his hosts wanted to take him, saw what they wanted him to see, and learned what they wanted to teach. A trip to the distant, beleaguered outpost at Dunkirk reflected France's concern for the embattled fortress, a matter of great national pride. Of the four air stations the French wished the United States to operate, Dunkirk ranked first. It was also the initial site visited.[13] Yet Dunkirk bore little relationship to Whiting's instructions from Navy headquarters. Similarly, his lengthy discussions with Captain Charles Lambe (RN), commander of Royal Naval Air Service operations in Dunkirk, highlighted the importance of anti-submarine operations in the region, a mission with which the Royal Navy had struggled for three years.

Though certainly the single most important source of overseas aviation information and recommendations, Whiting did not act alone, and a growing chorus of voices soon joined his. Marine Capt. Bernard L. Smith, an aviator for five years, had been serving at the United States embassy in Paris since August 1914 as an assistant to Naval Attaché Lt. Cdr. William R. Sayles. During that time he visited numerous French aerial units at the front and even made flights over enemy lines. In the fall of 1916, he carried out a secret mission to Switzerland in search of aeronautical data. Smith met the First Aeronautic Detachment when they reached France and accompanied Whiting on several early inspection trips. The following month, he returned to the United States to report on everything he had learned and personally delivered a proposal to establish aerial patrol stations on the coast of France. Assigned to Capt. Noble Irwin's growing aviation headquarters, he assisted in the areas of design, construction, testing, and acceptance of new combat machines.[14]

Attaché Sayles also dispatched information to Washington, especially after the arrival of the First Aeronautic Detachment, reporting on the progress and activities of Whiting and his men. Utilizing his considerable language skills, Sayles acted as go-between for Whiting and the French Ministry of Marine and during much of the summer of 1917 served as the principal conduit for information to the Navy Department. Sayles reported on the assignment of personnel to various French schools and forwarded requests that certain officers be dispatched to France, including Earl W. Spencer, Albert C. "Putty" Read, William M. Corry, and Harold T. Bartlett. He also recommended that Paymaster Omar Conger, who accompanied the First Aviation Detachment, be retained in France to carry out the various financial and contracting responsibilities necessary to put any American aviation program in place.[15]

Paymaster Conger, a member of the First Aeronautic Detachment contingent, represented the Bureau of Supplies and Accounts. He brought with him a hefty supply of gold coin and authority to conduct business negotiations and execute contracts. It was Conger who bore responsibility for feeding and sheltering the American aviators, arranging training regimens, and handling construction and logistical details for any programs and stations established in France. He communicated frequently with his boss, Paymaster General and Bureau of Supply Chief Rear Adm. Samuel McGowan. An early missive of June 18, 1917, offered a general description of conditions in Europe and a brief overview of a proposed aviation program. Many more letters and cables followed.[16]

While Whiting toured French bases in June and July, Capt. Richard H. Jackson reached France, assuming the role of senior naval officer in Paris and serving as Admiral Sims' envoy to the Ministry of Marine. An officer who obtained his first commission at the direction of Congress for heroism exhibited in 1889 during a storm while on board screw steamer *Trenton* in Samoa, he later held a variety of seagoing and shore-based posts, culminating with command of the battleship *Virginia*. At his new post in Paris, he forwarded reports on the progress of Whiting's negotiations, including descriptions of a tentative program to establish four American stations and a call for dispatch of eight hundred to one thousand additional men for the effort. Jackson also outlined supply and construction obstacles that might hinder such an effort, urged shipment of prefabricated barracks and hangars from the United States, and endorsed Attaché Sayles' call for various aviation officers.[17]

Bolling Aeronautical Commission members Cdr. George C. Westervelt and Lt. Warren G. "Gerry" Child, provided yet another source of information.

The commission headed by Major Raynal C. Bolling was sent by the War Department to study aircraft in use by the Allies in order to recommend which types should be placed into production in the United States and which should be purchased from European manufacturers. In addition to Bolling, who was General Counsel for the United States Steel Corporation in civilian life and had learned to fly in 1915 while a member of the New York National Guard, the commission included financiers, industrialists, engineers, and aviators, such as Maj. Edgar S. Gorrell, a veteran pilot of the 1st Aero Squadron who held a master's degree in aeronautical engineering from MIT.[18] The members of the commission, approximately one hundred in all, embarked for Europe June 17, 1917, and soon initiated a thorough canvas of production facilities through the Allied countries.[19]

In addition to their duties with the commission, Westervelt and Child also assisted naval aviation's fledging efforts in Paris and elsewhere. They both participated in a large July 8 conference at the Ministry of Marine that included Bolling, Whiting, Conger, and numerous high-ranking French officers. Discussions covered the availability of French aircraft for any proposed Navy stations, possible employment of American workers in local factories, along with the possible creation of United States manufacturing facilities in Europe. The visitors also received information regarding France's projected fifty-base expansion plan and a request that the United States operate twelve of the new naval air stations.[20] Westervelt and Child accompanied Whiting on some of his inspection tours, notably an August 7 trip to Dunkirk that included visits to outlying aerodromes, the enormous aviation depot at St. Pol, and a wireless tracking station. They also provided advice to Admiral Sims regarding material and to the Department concerning authorization of patrol stations. It was Westervelt in mid-August who recommended that Sims appoint a staff officer with the rank of captain to oversee aviation affairs in Europe.[21]

Even with this influx of officers and the increased flow of information back to the United States, Whiting remained the focal point of the Navy's nascent aviation efforts in France. Based on his discussions, research, and inspection tours, Whiting penned four lengthy reports/recommendations for policies to be pursued by the United States, including a detailed description of the organization/operation of a naval air station, the scope of the program necessary to defeat the U-boat menace, the state of development of the Allied naval aviation effort, the various ways naval aircraft might be employed, and current and planned anti-submarine tactics and initiatives. His reports provided a virtual encyclopedia of ideas and the most comprehensive overview available to planners in Washington. Out of these missives emerged the Navy's commitment in August and September to construct and operate fifteen anti-submarine patrol stations along the coast of France.

As part of his ongoing examination of aviation developments, Whiting visited a Royal Naval Air Service aerodrome in Coudekerque near Dunkirk in August 1917 and came away deeply impressed with the offensive capacity of the large Handley Page 0/100 aircraft stationed there. He soon recommended the Navy initiate a major bombing campaign, something not mentioned in his earlier communications. Whiting labeled attacks on enemy submarine pens in Belgium "indispensable" and proposed that a force of 500 fighters and bomber seaplanes be employed against German bases at Heligoland, Wilhelmshaven, Emden, Cuxhaven, and elsewhere.[22] He even advocated seizing bases in neutral Holland or Denmark to support this work. As an alternative, Whiting recommended converting 15–20 train ferries into seaplane ships to undertake such raids. In mid-September, he repeated his proposals, calling for an immediate decision, estimating six to nine months would be required to gather and train the proper force.[23]

In the United States, Cdr. Henry Mustin, one of the Navy's most outspoken aviation proponents, responded to a fleet-wide invitation from Secretary Daniels for innovative ideas for pursuing the naval war by proposing continuous bombing of military-industrial targets such as the submarine shops and facilities at Emden and Wilhelmshaven and related works at Essen. Torpedo attacks on German submarines and other ships in their North Sea bases would supplement these raids.[24]

At the end of the summer, when Admiral Mayo visited London, aviation had been one of the topics on his agenda. A British plan to construct additional patrol stations in Ireland was discussed briefly. Shortly after Mayo's trip, Cdr. (soon Capt.) Hutchinson I. "Hutch" Cone reached England to assume his new responsibilities as head of United States Naval Aviation Forces, Foreign Service (USNAFFS). He met with Admiral Sims, Lieutenant Whiting, and Paymaster Conger, and began familiarizing himself with the overall aviation situation in Europe. On October 2 he sat down for an extended discussion with the director of British naval aviation, Commodore Godfrey Paine, the Fifth Sea Lord. Paine returned to the question of Irish stations, revealing that the Admiralty wished the Americans to operate a series of patrol facilities guarding the approaches to the Irish Sea and Liverpool. In the next few days, Cone and his party conducted a whirlwind tour of British aviation sites, including a balloon school at Roehampton, the giant patrol base at Felixstowe, and an experimental station at Isle of Grain. A lengthy inspection trip to Ireland and Scotland followed. Ultimately, the United States agreed to operate four patrol stations and a kite balloon facility in Ireland, one small step toward implementing President Wilson's demands for more aggressive anti-submarine action.[25]

Of greater significance to the development of any future bombing programs, Cone's early days in Britain also resulted in a tentative Anglo-American agreement to undertake a large-scale, seaborne offensive against German fleet bases in the Heligoland Bight area utilizing long-range flying boats launched from specially designed lighters. Killingholme on the River Humber was selected as the likely site for the American portion of the effort.[26] This was the first of several bombing initiatives ultimately planned or undertaken by the Navy during World War I and indicated the service's budding commitment to a new and untried technology and strategy to defeat the submarine, not by chasing them "all over the farm," but by "going to the nest and crushing them."[27]

Despite its heretofore limited interest in aviation, the Navy Department eventually embraced many of these proposals, marshaling vast manpower and material resources for that purpose. In effect, discussions concerning the proper role for naval aviation became subsumed in the larger debate concerning the overall tactics to be employed against the German navy, and specifically its submarine arm. These discussions rested on the perception held in Washington, that British strategy and tactics had been unnecessarily passive.

CNO Benson's extended trip to Europe in November and December 1917, though not originally intended to focus on aviation issues, proved crucial to development of the Navy's bombardment program. During his weeks abroad, he met with many British and French officers and officials, as well as American personnel in London, Paris, Brest, and elsewhere. While on a visit to the Grand Fleet, Benson learned that the Royal Navy believed assaults against enemy submarine bases could not be mounted until the attacking force gained control of the air. Impressed with this logic, he wrote to Daniels November 14 emphasizing the need for a "sufficient number of vessels carrying aircraft to permit the extensive bombing operations absolutely essential to the successful attack on German fleet, munitions, and shore batteries."[28]

While in Paris, Benson held detailed discussions with Captain Cone concerning development of naval aviation in Europe. According to the headquarters log, the admiral visited the office November 23, the same day a meeting took place among commanders of American aviation stations then operating in France: Grattan C. Dichman from Moutchic, Godfrey de C. Chevalier Chevalier from Dunkirk, Harold Bartlett, in charge of Navy flying personnel detailed to the French gunnery school at Cazaux, John L. Callan, and Kenneth Whiting, along with Paymaster Omar Conger, the man responsible for equipping and paying for the entire effort. Presumably, Benson met each of these officers and received reports of the progress and difficulties encountered thus far. There is little doubt that Benson's extensive conversations with American flying officers in France and his visits to training and operational facilities played a major role in energizing his crucial support for major aviation initiatives in the coming weeks and months.

Three days later, Benson returned to aviation headquarters and, accompanied by Cone, traveled out to the sprawling French air station at Villacoublay, just outside Paris "to look at the flying." More was to follow. On December 5, Benson, Cone, Sims, and their staffs set off on an extended tour of American aviation facilities on the west coast of France, heading first for Naval Air Station (NAS) Le Croisic, guarding the approaches to the important port of St. Nazaire, the only station currently conducting antisubmarine patrols. The group then moved on to Moutchic, the Navy's rapidly expanding training facility situated on Lac Lacanau about thirty miles north of Bordeaux. The party returned to Paris December 9 "after a trip to the various stations on the west coast."[29]

Although aviation had not previously headed Benson's list of concerns, the subject arose frequently, whether in discussions with Admiral David R. Beatty, Admiral Ferdinand de Bon, or during several visits to Navy headquarters in Paris. A trip to Villacoublay and the December inspection tour of facilities along the French coast permitted him to observe flying operations firsthand. If there were any more aggressive and visionary officers in the U.S. Navy than the aviators, they would have been hard to identify. He must have received both an eye- and earful. Despite Benson's well-deserved reputation for conservatism and his obvious disdain for the institutional objectives of naval aviators, he nonetheless occupied a pivotal position in Washington and his growing support for aviation as an offensive tool proved crucial in later policy decisions establishing the strategic bombing program.

Benson's discussions with both Allied and American officers clearly opened his eyes on the subject. When he returned to Washington, he reported to Daniels and the President, saying, "Offensive operations in the air I consider a necessary preliminary to other forms of naval offensives against enemy bases."[30] His biographer Mary Klachko observed, "After his European trip, Benson's interest in naval aviation became very apparent."[31] Adm. David Taylor, the air-minded chief of the Bureau of Construction and Repair, concurred, observing that "a flurry on the aeroplane situation" followed the CNO's return (to Washington). This included doubling construction of single-engine flying boats and quadrupling Cone's request for twin-engine flying boats. Briefed on the proposed Killingholme bombing initiative, Benson "approved immediately," sending orders to Washington to begin procuring aircraft, engines, and lighters. Kenneth Whiting received instructions to return to the United States as head of the project. Capt. Noble Irwin, leader of naval aviation in Washington, was directed to complete air stations in France as quickly as possible.[32]

Word of Benson's "conversion" quickly crossed the Atlantic. In January, Cone received word of the shift in attitude via letter from an old Annapolis classmate then serving as Secretary of the General Board. The CNO, he reported, had recently addressed the regular monthly meeting of the General Board and expounded at length on "the great work that is being done in establishing aviation stations and the work in connection therewith."[33] In London, Sims, who strongly supported aviation initiatives and personally selected Cone to command such operations, also sensed the change in emphasis. "As far as I can see," he noted in a January 29, 1918, letter to Capt. Mark L. Bristol, former head of naval aviation (1914–16), "the principal dignitaries at home are absolutely alive to the necessity of giving all practicable assistance, and I believe they are doing this to the extreme of the extent of their power. This is a pretty strong statement, but I believe it to be true." It seems that Benson's long-held reputation as an enemy of naval aviation requires a certain amount of modification.[34]

And Benson was not alone in these attitudes. In fact, several officials ultimately came to see aviation—only recently the Navy's unappreciated stepchild—as a formidable weapon with which to challenge the German fleet. For flyers and their supporters, going on the attack would demonstrate aviation's value, both within the fleet and vis-à-vis the U.S. Army. Americans seemed particularly inspired by British aviation and tended to model their actions after those of the innovative and offensive-minded Royal Naval Air Service, the world's largest force. It was no coincidence that the General Board, the Marine Corps, the London Planning Section of Sims' headquarters, and naval aviation headquarters in Paris all turned their energies toward an expanded anti–submarine offensive within weeks of Benson's return to Washington. Among the possibilities they explored was offensive (strategic) bombing, something heretofore outside the Navy's realm of interest.

Even as attitudes toward employing aviation in an offensive mode gained currency, the foreign base construction program authorized in August and September 1917 to protect American and Allied shipping moved ahead. By the end of the year, the Navy had made significant progress inaugurating a large program designed to place anti-submarine aviation assets along the coasts of France, England, and Ireland, as well as the United States. Based on Whiting's recommendations, fifteen patrol, training, or supply bases were being built or planned in France, designed to protect convoys headed toward the principal disembarkation ports of Bordeaux, St. Nazaire, and Brest, as well as the hotly contested waters off Flanders. In Ireland, the United States committed to manning four large patrol stations guarding the northern and southern entrances to the Irish Sea. On the eastern coast of England, the Navy planned to operate its largest patrol base at Killingholme, combining patrol/convoy escort duties over the North Sea and a visionary scheme to attack the German fleet via flying boat bombers carried by special lighters. None of these initiatives, however, directly targeted German submarine bases located at Bruges, Ostend, and Zeebrugge, the true hornets' nest. Events unfolding at the Navy's new, exposed station at Dunkirk would soon change this situation.

Chapter Four
The Dunkirk Dilemma

E
fforts to establish naval air stations in France and Ireland encountered numerous obstacles, including bad weather, inaccessible or inconvenient locations, and sometimes crippling shortages of labor and supplies. Many of these conditions also prevailed at the new NAS Dunkirk in Flanders. Having chosen this coastal outpost as a site for aviation activities, the Navy soon faced the harsh realities entailed in that decision. Dunkirk quickly proved to be a singularly problematic place from which to conduct seaplane operations. Located only a few miles behind the front, it endured continuous bombardment by enemy land, air, and naval forces. Overwater aerial patrols often fell victim to land-based German pursuers or more advanced seaplane fighters. Weather proved abominable. The narrow harbor, studded with obstacles and vessels, foreclosed the possibility of employing larger, more capable aircraft. Allied patrols rarely encountered enemy U-boats.

These difficulties played directly into the eventual decision to create the Northern Bombing Group. As the months passed, it seemed to many that NAS Dunkirk could never carry out its intended mission of crippling the enemy's submarine force operating from Bruges, Zeebrugge, and Ostend. The British abandoned seaplane operations there in favor of land-based aircraft, mounting day and night bombing raids against German submarine bases instead. Was this a better alternative for the Navy? Would such a shift enrage the Army, or the brass in Washington?

In a very real sense, the eventual decision to create a Northern Bombing Group to mount round-the-clock aerial attacks against Germany's extensive

U-boat complex in Belgium evolved from the early activities and recommendations of Lt. Kenneth Whiting. Accompanied by Paymaster Omar Conger and Captain Bernard Smith, his initial inspection tour, and in some ways his most important, took him to Dunkirk on June 19, 1917. There he examined the French seaplane station located on the eastern edge of the narrow, crowded harbor. Next day he met with Captain Charles Lambe, commander of RNAS squadrons in the region. The Americans received a detailed briefing of conditions in the area as well as a tour of the British seaplane base located adjacent to the French facility.[1] Whiting also inspected the site (chosen by the French) for a proposed American station and the enormous Allied aviation installation at St. Pol, but lack of time precluded a visit to the British bombing aerodrome at nearby Coudekerque. The energetic American officer returned to Paris on June 21.[2] More conferences and inspection tours followed quickly, resulting in a recommendation cabled to the Navy Department in early July that the United States establish four stations on the French coast, including Dunkirk.[3]

The irony of Whiting's recommendation and the Department's acceptance of his proposal regarding Dunkirk lay in the obvious fact that operations there had little to do with the Navy's principal responsibility of moving ships, men, and supplies safely to Europe. This, after all, was the justification for authorizing establishment of air stations at places like Le Croisic, St. Trojan, and Brest, all situated to protect access to France's west coast ports, or the later mandate to build four large bases in Ireland to patrol the sea lanes to Liverpool. By contrast, the original mission envisioned for a station at Dunkirk would have been to harass U-boats as they worked their way into the North Sea and the English Channel, an area destined never to see an American troop transport. It was, however, in line with the concept of going after hornets in their nest.

Whiting returned to Dunkirk in early August, this time accompanied by Commander Westervelt and Lieutenant Child, then touring Europe investigating Allied aviation programs as part of the Bolling Commission. They wished to see what progress, if any, had been made at the new Dunkirk site and received assurances from the French that work would be completed by September 1. The aviators met again with Captain Lambe, visited several aerodromes and a wireless tracking station, and returned to the aircraft depot at St. Pol. Whiting also took time to inspect two bombing squadrons led by Wing Commander Spenser Grey, one operating huge Handley Page aircraft and the other employing Airco DH-4 day bombers.[4] Whiting took particular delight in viewing the beached wreck of *UC-52*, a German mine-laying submarine.[5]

By recommending establishment of an aviation outpost at Dunkirk, Whiting injected the U.S. Navy deep into the realm of the Royal Naval Air Service, largest and most aggressive of all European naval air services. In becoming de facto tactical partners with the British, he virtually determined the Navy would absorb many of their priorities, strategies, and ideas, including the use of naval air power against land targets. Even before a single bluejacket reached Dunkirk, Whiting envisioned employing naval aviation in an offensive, bombing mode to attack submarines, their bases in Belgium, and other German targets far beyond the immediate battlefield. This fit entirely with the American administration's oft-expressed view that naval assets should go in harm's way in their anti-submarine campaign. Finally, by choosing the thoroughly inadequate Dunkirk location as the United States' primary base of operations, Whiting inadvertently ensured that the very mission pursued could never be accomplished with the seaplane tools available, leading almost inevitably in the direction of large, land-based bombers instead.

Many problems later encountered at Dunkirk were well-known or foreseeable. Rubble and shell-holes dotted the urban landscape. During his initial visit, Whiting learned the city was under constant bombardment by the enemy. Heavily built bombproofs were necessary to protect personnel at the Allied air stations. An extreme range of tide made handling aircraft difficult. Morale had plummeted. Six French aircraft were lost on a single day in April. On the day in June when Whiting first arrived, the British lost two scouts, a bomber, and a small motor torpedo boat. During a return visit in August, the RNAS announced they were abandoning seaplane patrols in favor of land-based scouts.[6] A midday bombing raid resulted in fifteen civilian deaths, with seventy-five more injured, mostly Chinese laborers that had been brought in for the construction work. Whiting also experienced "heavy rain falling almost incessantly," a portent of things to come.[7]

Right from the beginning, construction at the American station fell behind schedule and the work of civilian contractors, Moroccan laborers, and local soldiers proceeded very slowly. By mid-1917, France suffered severe shortages of labor and building materials, greatly hindering efforts to erect hangars, shops, barracks, and offices. The promised September 1 completion date for the American station came and went. An initial dirt-floored barrack was completed September 2 and the first canvas hangar

October 5. By the end of the next week, six barracks had been closed in. Excavation of a bombproof shelter began a few days later, but construction lagged due to shortages of materials for walls and roofs. Galleys remained unbuilt, as did latrines. Sluggish progress continued through the fall and early winter. Only the arrival of bluejacket labor from the United States would hasten the process.

The influx of naval personnel proceeded very slowly, however, and those who reached the beleaguered port city often found limited or nonexistent accommodations. The first American staff on the site included Ens. Clarence R. Johnson, tasked with overseeing the operation. A group of 50 seamen arrived in early November. With no place to mess or sleep, they ate at a local restaurant and slept in an old city building, with blankets and mattresses secured from the French. By early January 1918, the enlisted contingent had risen to 129, still well short of the 200-man complement originally envisioned. Of this group, 90 lived at the station, while the remainder continued to board in the city.[8]

Pioneer naval aviator Lt. Godfrey de C. Chevalier commanded the struggling outpost. A native of Providence, Rhode Island, he graduated from the Naval Academy in 1910 where he excelled in track and field, though not in discipline. Described as "small, unusually handsome," he might never have graduated at all but for his extraordinary charm, which made everyone who knew him want to excuse his offenses.[9] Chevalier began aviation training at the Annapolis Aviation Camp in October 1912 and completed instruction the following year, eventually earning the designation Naval Aviator No. 7. Several years of busy flight duty followed. "Chevy," as he was known to his friends, also possessed a reputation as an officer who enjoyed a good drink. A member of the officer contingent of the First Aeronautic Detachment, Chevalier passed through the entire regimen of French aviation instruction, with stops at Tours, Avord, Pau, and Cazaux, making him the best-trained pilot in the U.S. Navy. He was twenty-eight years old when he arrived to take up his duties as commander of NAS Dunkirk in November 1917.[10]

Chevalier immediately encountered an entire menu of challenges. In addition to construction delays and personnel shortages, he discovered that delivery of necessary aircraft and armament also lagged woefully behind schedule. A few unassembled machines arrived in late October and early November. For fuel, oil, lubricants, and similar necessities, the Navy depended on the RNAS. More aircraft arrived in the next few months, though often without armament or other vital equipment. The first test flights did not occur until late January. Even had aircraft and equipment been available, NAS Dunkirk remained unready to begin antisubmarine activities for one very basic reason—there were few, if any, pilots or observers.

At the time it established the station, the Navy had virtually no pilots trained to fly the aircraft and conduct the missions anticipated for the base. Enlisted student aviators who traveled to France with the First Aeronautic Detachment followed a lengthy instruction regimen that took them to Tours, St. Raphael, Moutchic, Issoudun, and Ayr. None would be ready to carry out combat activities until late March. Young college boys, members of the First Yale Unit, also faced lengthy delays making their way to the front, including stops at Lac Hourtin, Moutchic, Gosport, Turnberry, and Ayr. They, too, would not reach northeast France until late February or March.[11]

Flying personnel at the station were thus limited to Commanding Officer Chevalier, Lt. Virgil C. "Squash" Griffin on loan from Paris headquarters, and Lt. (jg) Artemus L. "Di" Gates, a member of the First Yale Unit, but without combat experience. A few former members of the Lafayette Flying Corps had enlisted in the Navy and would eventually reach Dunkirk as well. The same situation held true for the enlisted observers/gunners/bombers, without whose expertise anti-submarine missions could not be mounted. Consisting almost entirely of members of the First Aeronautic Detachment, these men trained at St. Raphael and Moutchic, and then received further instruction at Cranwell, Eastchurch, and Leysdown in England. They began reaching Dunkirk in March 1918.[12]

Shortages and delays greatly hampered initiation of anti-submarine activities, but nothing compared to the nearly daily assault carried out against the city and its military facilities by the German army, navy, and air force. Of all the American naval air stations established in Europe, more than twenty-five in all, only Dunkirk endured nearly continuous attacks from enemy aircraft, naval vessels, and long-range artillery. Whiting witnessed a heavy air raid during a visit in early August. In September, British repair and machine shops were "almost bombed off the map and had to be moved to a new and less exposed position."[13] On October 18, a flotilla of enemy destroyers attacked the harbor, firing 250 shells and damaging a monitor with three torpedo hits. A report to headquarters in November declared that RNAS St. Pol was "practically wiped out by German bombs and gunfire."[14]

A few weeks later Whiting noted, "The bombing of Dunkirk has become a very serious question and until now no reliable means have been found to stop the German bombing planes operating at night."[15] Just before Christmas, a raid against the French patrol base proved the worst yet, setting a seaplane on fire and burning three or four hangars. The American station weekly report of January 23, 1918, stated, "Nightly raids by enemy aeroplanes rendered construction and assembly work difficult. Work is progressing as well as can be expected, however, under the circumstances. All personnel are obliged to live in dugouts as protection from enemy bombs." In July 1918, the *New York Times* reported that Dunkirk had endured 157 attacks.[16]

At a more personal level, Machinist Mate Irving Sheely spent his initial days in Dunkirk inspecting areas of the city that had been bombed or shelled. Then he went to work erecting hangars and constructing a dugout. In one letter home, he reported that he would soon be heading to England for additional training, "just till our station gets built up again. You would laugh to see me run for a dugout when the bombs start bursting around us." A few weeks later, enjoying a respite from bombardment while undergoing training at RNAS Cranwell in Lincolnshire, Sheely added, "I don't like to go back to Dunkirk. It's a rotten place and the Huns bomb it every moonlight night, so you see that sometimes you don't get very much sleep. Besides, it is not very pleasant to be standing in the dugout half-dressed either." Miraculously, all the bombs and shells failed to kill a single American in the year they occupied the site.[17]

Along with the incessant deluge of bombs and artillery shells came the dreadful weather that blanketed the entire region from September through April. Machinist Mate Alonzo Hildreth who reached the station in November kept a detailed diary of events transpiring during the winter of 1917–18 and frequently cited the cold, wind, rain, and snow. In the best of circumstances winter weather on the Channel coast varied from dreadful to abominable. To make matters worse, the winter of 1917–18 was one of the coldest ever recorded.[18]

In March 1918, after many months of delays, the essential pieces finally began coming together. A limited supply of aircraft was on hand and ready to carry out operations. A small cadre of pilots, observers, and mechanics had been assembled. Several test flights were completed. Although the station remained well understrength in terms of personnel and equipment, there were enough of each to begin. And then on March 21 the German Army launched its great attack on the Western Front, the *Kaiserschlact*, designed to end the war before the American Expeditionary Force's growing strength tipped the balance against them. Operating cheek-by-jowl with the RNAS, the Navy offered some of its limited Dunkirk manpower to fill yawning gaps in British squadrons, including many of its best pilots and observers. Anti-submarine missions would have to wait.[19]

The return of personnel from British units in late April, coinciding with the arrival of marginally better weather, allowed more active flying to commence, and with it a full comprehension of how treacherous it would be to conduct large-scale operations from the site the Navy had labored so hard and long to complete. A single crane on the quay made launching and retrieving aircraft a tedious affair. Gates, the station's chief pilot, recalled that, "The location of the station proved especially poor owing to the great difficulties getting in and out of the small harbor with a seaplane." Cramped conditions frequently necessitated "flying over the city to make a landing or a getaway, we suffered numerous losses from crashes."[20]

If the Navy needed any proof Dunkirk harbor might not be suitable for supporting the important anti-submarine mission, a continuous string of mishaps experienced by pilots and observers quickly gave substance to the many warnings received previously. On February 26, 1918, Ens. Curtis S. Read and his observer, Edward "Eich" Eichelberger, embarked on a training flight in a 200-hp Donnet-Denhaut flying boat. They managed to lift off from the crowded harbor, but twenty minutes later spun into the water. Both men were killed. Shortly after arriving in late March, Ens. George Moseley, a Lafayette Flying Corps combat veteran, wrote home describing conditions there. "The harbor here . . . is very narrow and the wharfs on each side are lined with ships of all descriptions . . . it is almost impossible to get into the air when the wind is blowing across it."[21]

On a blustery day in early April, Moseley climbed into the cockpit of his Hanriot-Dupont scout and taxied out into the harbor. Opening the throttle, he accelerated rapidly until he began to rise from the water. A sudden gust of wind lifted his right wing, forcing the left pontoon into the water, slewing the aircraft to the left, now heading directly toward a British destroyer. Attempts to turn proved fruitless so Moseley endeavored to leap over the vessel. He yanked back on the stick, raced the motor, jumped into the air and hit the ship's mast, sending his machine nose-first onto the destroyer's deck. Although he walked away from the crash, he remembered nothing and awoke in the hospital with a doctor stitching his forehead. Moseley encouraged the medical man, saying, "That's the stuff, Doc; stick it in, boy, stick it in."[22]

The parade of accidents continued. On May 5, Ens. Kenneth MacLeish, also in a Hanriot-Dupont scout and riding a tail wind, overshot his landing and crashed in the harbor. He just missed smashing into the seawall, but broke both of the aircraft's pontoons and one wing. Two weeks later, enlisted pilot John Ganster made a turn while above the city, lost control of his machine, and crashed into the roof of a house on Rue Carnot. He was thrown clear of the wreckage, but died a short while later. Following the war, the homeowner entered a claim to the United States government for reimbursement for the damages caused by the accident. Within weeks, pilot Djalma Marshburn died in an accident as well.[23]

All these negative factors—bad weather, high tides, inadequate room to maneuver, innumerable obstacles both fixed ashore and afloat, obsolescent equipment, ceaseless German bombardment, stout enemy aerial opposition, ineffective tactics—combined to convince many observers and planners that NAS Dunkirk could never carry out its intended mission. As early as January 1918, there were suggestions the effort be cancelled or scaled back. In February, Admiral Sims' London Planning Section reviewed the overall aerial anti-submarine effort and enumerated the many challenges it faced. As to the base at Dunkirk they stated, "The station . . . appears unnecessarily close to enemy lines, and, as the British have already been compelled to abandon their station there, we suggest the advisability of further consideration of this location with a view to its possible abandonment in favor of a position somewhat more retired." The same group strongly recommended the Navy abandon use of flying boats as bombers against enemy bases and fighters over narrow waters.[24]

Captain Cone agreed with the latter proposal. Concerning the viability of the station at Dunkirk, he also concurred with the board's assessment, but claimed no alternate site existed along the French coast. Additional comments by the British Plans Division recommended that "any further American naval aircraft can best be devoted to offensive operations against the enemy submarine bases." Seizing control of the area and launching major attacks against the submarine infrastructure in Flanders would lead to "the enemy being compelled to suspend his operations from the Belgian coast. This would be of great advantage." U-boats would be forced back to the Heligoland Bight where their effectiveness would be reduced "to the extent of at least 50 percent."[25]

All the information gathered by the board reinforced the contention that only land-based aircraft could accomplish the objective of seriously impairing submarine operations in the region. Patrolling and occasional attacks at sea simply would not suffice. The British Plans Division argued strongly for the use of heavy bombers instead, if allowed to concentrate against U-boat bases. They noted that large, multiengine Handley Page aircraft first reached the front in November 1916 but were sent to eastern France "where they were of little value." During the following year there were "long periods during which bombardment of the submarine objectives on this coast have been suspended by order," with aircraft given targets of "military importance" rather than "naval objectives." Many RNAS officers, "particularly those best qualified to judge," believed submarine activity in the region "might have been adversely affected in no small measure if the whole naval bombing resources available in 1917 had been devoted to this single purpose."[26]

The Navy's frustrating experience at NAS Dunkirk in 1917–18 laid the groundwork for the later emergence of a strategic bombing program. Indeed, Clifford Lord, the first and still finest chronicler of naval aviation's early decades, called Dunkirk the grandfather of the Northern Bombing Group.[27] Living and working as they did in an active combat zone, station personnel became intimately familiar with the devastating impact aerial bombardment could have. On clear nights, they listened for the droning of Gotha bombers and awaited the wail of "Mournful Mary," the city's warning siren. Anti-aircraft fire lit up the evening sky while raining down torrents of shrapnel. With British aerodromes located just a few miles away, Americans could watch allied bombers in action, meet with pilots and crew, and discuss strategy and tactics. Captain Lambe provided constant encouragement, forcefully advocating expansion of Navy bombing efforts.

At the same time, operational difficulties encountered at Dunkirk inevitably dampened enthusiasm for seaplane missions attempted there. Though the Navy did not abandon Dunkirk as an operating base in 1918, a combination of factors caused planners to downgrade their expectations of what could be accomplished to suppress enemy submarine activities and reevaluate how much manpower and equipment should be allocated to the site. If the U-boat could not be defeated by flying boats operating out of Dunkirk, then alternative methods would be necessary, methods that would take the Navy far from its initial reliance on seaplane patrol and into the new field of strategic bombing.

Chapter Five
Bombardment Aviation

America to the Rescue

The Navy's aeronautic efforts at Dunkirk and elsewhere in Europe did not occur in a vacuum, but instead formed one piece of the overall mosaic of Allied aviation plans, initiatives that included French, British, and U.S. Army Air Service programs to harness the destructive potential of strategic bombing. In time, these labors intertwined with Navy priorities.

When the United States entered the war, French bomber aviation was plagued by poor material. It had no satisfactory aircraft and had to rely on obsolete, inadequate, or mediocre machines. France's newest airplane, the Voisin 8, introduced toward the end of 1916, carried less than a quarter of the ordnance load of Italy's Caproni or Britain's Handley Page bomber.[1] Despite the disbanding of Britain's No. 3 Naval Wing in spring 1917, the enthusiasm of the French Air Service for long-range bombing persisted. In April, Captain Henri de Kérillis regretted the feebleness of his country's strategic aviation and longed for reprisal raids to "strike the morale of the enemy, to intimidate him."[2] Kérillis believed that a fifty-plane raid on Munich "would have flung enough German entrails on the pavement of the city to give the torpedoers of the *Lusitania* and the arsonists of Reims pause for reflection."[3]

Kérillis, an expert on day bombardment, had commanded the Sixty-sixth Bombing Escadrille, which carried out all of the daylight raids in Lorraine, from March 2 to the end of September 1916. This elite squadron was equipped with the Caudron G.4, an aircraft that was not very fast or maneuverable, but climbed quickly. Kérillis insisted that his

pilots use rigorous discipline while flying and invented the tactical formation called "flight of ducks," which allowed for flanking fire and mutual
defense. He made sure that his crews flew above 10,000 feet where it was
difficult for German aircraft to maneuver, always took a different route,
varied their schedule, avoided enemy airfields, and crossed the front lines
at high altitude.[4]

The French, eager to extend their long-range bombing program, looked
to America to solve their material problems. One month after the United
States declared war on Germany, the French general staff drafted a plan for
U.S. participation in the air war issued under the signature of Gen. Henri
Philippe Pétain, commander in chief of the French Armies of the North and
Northeast. In a memorandum to the Minister of War on May 6, Pétain suggested that the ideal American aerial contribution would be to provide 30
pursuit groups and 30 bomber groups, each group to comprise 6 squadrons
of 12 aircraft for a total of 4,320 machines. The number of aircraft, rounded
up to 4,500, became the basis for the figures submitted to the United States
in the famed Ribot Cable that laid the foundation for the airplane building
program initiated in America.[5] The French request, according to Maj. William Mitchell, was for strategical aviation units separate from those directly
attached to army units.[6] The latter, which Mitchell termed "Tactical Aviation," would ensure observation for fire and control of our own artillery.
"Strategical Aviation" would be used to attack enemy material of all kinds
behind the lines. To be successful, it needed to be composed of large groups
of airplanes organized separately from those directly attached to army units.[7]

Mitchell, assigned to the Aviation Section of the Signal Corps, had been
sent to France before war was declared to observe the air war being waged in
Europe. Although he had earlier taken flying lessons at his own expense, he
was not a qualified pilot. When Mitchell arrived in Paris on April 17, 1917,
he immediately joined the American military mission and quickly set up
an unofficial aviation office in a room in the Paris branch of the American
Radiator Company staffed by a group of expatriates and French volunteers.[8]

Within weeks of his arrival, Mitchell began making trips to the front. On
April 24, 1917, he became the first U.S. Army officer to view the war from the
air while flying in the rear seat of a French observation plane.[9] "One flight
over the lines," he later wrote, "gave me a much clearer impression of how
the armies were laid out than any amount of traveling around the ground."[10]
Mitchell quickly concluded that for a force to be able to operate with full

freedom on the ground, control of the airspace above was essential: both as a means of protecting the observers who performed the critical functions of aerial reconnaissance and artillery spotting, and as a defense against enemy strafing and bombing attacks. Reconnaissance, artillery spotting, and aerial photography were powerful new tools that had become essential for success on the battlefield; but securing air superiority was the mission that made all other operations possible.

Between trips to the front and his office back in Paris, Mitchell often stayed at the French headquarters at Cahlons where he was fortunate to meet General Pétain. As Mitchell noted in his diary, he was struck by Pétain's "forcefulness and positive way."[11] Although French air doctrine emphasized the tactical use of aviation, the general and other high-ranking officers were coming to understand that aviation could be decisive in an Allied offensive if it were used to attack enemy lines of communication. If the Allies developed aviation to the maximum, mused Pétain, "One can wonder if it would not become a decisive arm . . . in rendering the adversary blind, in paralyzing his communications and in demolishing his morale."[12] In the months ahead, Mitchell would come to rely on support of French aviation units specializing in strategic aviation.

Mitchell's freewheeling role as head of the U.S. Army's aviation effort in France came to an abrupt end on June 13, 1917, when Maj. Gen. John J. "Black Jack" Pershing arrived in Paris with his staff. Mitchell was among those on hand to greet the commander of the American Expeditionary Forces (AEF) when he reached the Gare du Nord Station. He informed Pershing's Chief of Staff, Lt. Col. James G. Harbord, that he was ready to proceed with any project the general wanted and presented Harbord with two papers that he had drafted dealing with American air policy and organization.[13] Although Mitchell's small office was absorbed into AEF headquarters, he was appointed to the Aviation Board established by Pershing to formulate an aviation plan for the AEF in France. As its senior airman, Mitchell, now a lieutenant colonel, was instrumental in preparing recommendations for an air service passed on to Pershing via Harbord, the board's chairman,

Staffing assignments and decisions concerning aviation affairs within Pershing's headquarters became more complicated on July 2 with the unexpected arrival of the Bolling Aeronautical Commission. The Bolling Mission—as it was often termed—landed at Liverpool on June 26, 1917, and conferred with officials in England before traveling to France. After consulting with local aeronautical experts, Bolling and key members of his staff left for Italy where they inspected all their aircraft factories, went through their technical section, conferred with manufacturers, visited the front, observed the use made of aviation, and talked to the men flying combat. The Italians, Gorrell later remarked, "treated them like princes."[14] The Mission quickly discovered that Italy was the only Allied country that had a functioning strategic bombing program and the Americans were impressed by the scope and success of the campaign, which had managed to field as many as 250 Caproni bombers at one time.[15]

When Bolling and members of his staff returned to Paris on July 27, 1917, they had visited all of the Allies and were ready to make recommendations.[16] These were summarized in a cablegram sent to Washington on July 30. The aircraft they recommended for production in the United States included three British planes: the Bristol Scout with an 80-hp Le Rhone engine for advanced training, the two-seat Bristol fighter with a 200-hp Hispano engine, and the Airco DH-4 for long-range reconnaissance and day bombing (with Rolls-Royce or other suitable engine); two versions of the French SPAD (one with a fixed 200-hp Hispano engine and the other with a 156 Gnome rotary engine); and the Italian Caproni with three 270-hp Isotta Fraschini engines or their equivalent for long-range night bombing.[17]

Sometime after the report reached Washington, Bolling wrote to Howard Coffin, chairman of the Aircraft Board, emphasizing the need for the United States to produce large numbers of bombers. "I cannot strongly enough express my concurrence in your view that our great effort should be directed toward an enormous quantity production of bombing machines both for day and night."[18] Unfortunately, the aircraft needed to create a strategic force were listed in third place when the official report of the Bolling Mission was published. Authorities in the United States interpreted this to mean that they were third in relative order of importance. As a consequence bombardment machines were not included in the initial manufacturing program. As will be seen, this was not the only cause for delays in production.

Instead of returning to the United States at the conclusion of their mission as originally planned, Bolling and some of his staff were absorbed into AEF headquarters. Mitchell, having taken over as Pershing's staff officer for air, proved unhappy with this arrangement, which created something of a dilemma for the general, who now faced the problem of what to do with two aviators of equal rank. To make room for Bolling, Mitchell was placed in command of all aviation activities in the Zone of Advance and Bolling all activities in the Zone of the Interior. In essence, Mitchell commanded the operational forces, while Bolling took charge of training and supply. This arrangement precluded unity of command within the AEF Air Service so that all disputes—of which there were many—needed to be referred to the chief of staff, or to Pershing himself if Harbord could not resolve them. To fix this clumsy arrangement Pershing appointed Brig. Gen. William L. Kenly as the chief aviation officer of the AEF on September 3, 1917.[19]

While the Americans worked out the kinks in the command structure of their Air Service and the types of aircraft to be produced, the British struggled with their own problems and the best policy to defeat the Germans. The zeppelins had been successfully countered by a series of defensive measures that exacted a steady toll on the airships, which were no longer considered a serious threat. To intercept these lumbering behemoths, the RFC stationed twelve squadrons containing 110 aircraft at strategic points along the English coast.[20] Newer aircraft equipped with sufficient climbing power to meet oncoming attackers and armed with incendiary bullets, were proving to be an effective means of bringing down the raiders.[21] Numerous anti-aircraft guns and searchlight batteries manned by over 17,000 servicemen aided the aerial defenders.

Having lost faith in the airships' ability to carry the war to Britain, the Germans turned to the Gotha heavy bomber that the firm of Gothaer Waggonfabrik had been working to perfect for the past year and a half. The G IV model, which entered service in the spring of 1917, was the first long-range German bomber to go into production.[22] With a wingspan of seventy four feet and powered by two 260-hp Mercedes engines, they had the range to reach London with a 400-kg (880-lb) bomb load and return to bases in occupied Belgium.

The Gothas conducted their first daylight assault on London on June 13, 1917, when fourteen of the heavy bombers attacked the eastern part of the city near Liverpool Street Station. The raid, which caused the heaviest casualties of any attack on England during the entire war, killed 162 people and injured 432. This unexpected development triggered consternation within the War Cabinet. In response, they decided to increase significantly the size of the air services and began drafting various schemes for its implementation.[23]

Several days after the raid occurred, Trenchard, who had accompanied General Haig to England to discuss the proposed ground offensive in Flanders, appeared before the War Cabinet to provide his ideas on combating this new aerial threat.[24] Trenchard's unrelenting faith in maintaining constant offensive pressure on the enemy and his insistence on keeping aviation tied to the ground plan dictated his response.[25] To halt the Gothas, Trenchard advocated seizure of the Belgian coast. If this could be achieved, the enemy bombers would have to cross over territory controlled by the Allies on both outward and return flights, thus making their operations extremely hazardous.

The next best solution, according to Trenchard, would be to intensify attacks upon all German air units and their bases. He believed that the attrition would eventually force the enemy aviation service to redeploy squadrons in order to minimize high tactical losses, eliminating any surplus machines that could be used to undertake long-range raids. Projecting his own ideas on strategic bombing, Trenchard assumed that the Germans would consider long-range aerial assaults only if they possessed air forces surplus to the requirements for tactical support of their own armies.

The Gothas reappeared over London on July 7, dropping bombs within the city proper and on the East End. Casualties included 54 killed and 190 wounded. Although 95 naval and military aircraft took off in pursuit, only one Gotha was destroyed. The War Cabinet had already approved a plan to increase the strength of the RFC from 108 to 200 squadrons, but the second raid, as Air Force historian George K. Williams suggests, hastened the review of the entire British air program "transforming it from an inter-service issue into a political question of public interest."[26] The cabinet believed that a prompt reprisal in kind against German cities was needed to dissuade the enemy from further attacks on the British citizenry now that civilians at home were sharing the same risks as their uniformed colleagues in France.

On July 11, 1917, the cabinet decided to set up a committee to examine the arrangements for home defense against air raids and the air organization in general. Prime Minister David Lloyd-George appointed himself chairman. Lt. Gen. Jan Christiaan Smuts, the only other member of the committee, did all of the work. Within eight days of the War Cabinet's mandate, Smuts had analyzed the problem of air defense and reported on the forces necessary to protect London. He then proceeded to investigate air policy and organization. Lacking any experience in aerial warfare, Smuts sought advice from the president of the Air Board, Lord Cowdray, and Sir David Henderson, director-general of Military Aeronautics. Smuts was unable to confer with either Haig or Trenchard, both of whom where in France preoccupied with the ongoing Ypres offensive. Thus, he failed to tap "the empirical wisdom of the air establishment in the field."[27]

By the time Smuts issued his second report on August 17, he had come to believe that the long-range bomber could change the nature of warfare and that there was no justification for continuing to subordinate the air services to the needs of the Navy and the Army. An air service he wrote "can be used as an independent means of war operations. Nobody that witnessed the attack on London on 11th [sic] July could have any doubt on that point. Unlike artillery, an air fleet can conduct extensive operations far from, and independently of, both Army and Navy. As far as can at present be foreseen, there is absolutely no limit to the scale of its future independent war use."[28]

Smuts grasped that the side that achieved superiority in industrial production and exploited that advantage would almost certainly win the war. Arms and machines, and not manpower, were now the decisive factors for victory. It took little imagination, he wrote, to realize that next summer while our forces on the Western Front are moving forward at a snail's pace, the air battle front will be far behind the Rhine, and the destruction inflicted on the enemy's centers of industrial production may well become a major factor in bringing about peace.[29] The War Cabinet endorsed Smuts' second report on August 24, and charged him with carrying out the recommendations contained within. According to one well-regarded source, Smuts' report laid the groundwork for the creation of an independent air service and the formation of an air policy that was to be the basis for all RAF thought for the next twenty-five years.[30]

After the second Smuts report was issued, the Air Board, via Commander Arthur V. Vyvyan, RN, asked Lieutenant Commander Lord Tiverton, RNVR, to prepare a paper that could be used to formulate a strategic bombing policy.[31] The Air Board—authorized by Parliament in December 1916—had been established to make recommendations concerning the types of aircraft required by the two services and to organize and coordinate the supply of material to obviate competition between the Army and the Navy.

Lord Tiverton,[32] a barrister by profession, had been studying aerial bombing since 1915 and was considered the Royal Navy's leading expert on bombs, bombsights, and bombing—skills he had mastered while serving first as armament training officer in the Air Department of the Admiralty, then in No. 3 Wing as its armament officer. Neville Jones later wrote that Tiverton "possessed a considerable knowledge of mathematics and scientific principles and this, together with an imaginative and enquiring [sic] mind, enabled him to undertake a detailed and systematic study of strategic bombing and the problems which it involved."[33]

Tiverton prepared his report in Paris, where he served as technical liaison officer in the Admiralty Air Department office. Not surprisingly, his findings echoed French bombardment plans and the lessons he had learned from his own service with No. 3 Wing.[34] Tiverton's analysis of the bombing problem contained four broad categories, starting first with the objective: the destruction of the German munitions industry. To accomplish this, he identified three industrial sectors located in specific geographical areas: the chemical group in Mannheim, the machine shop group in Düsseldorf and Cologne, and the raw steel group in the Saar Valley. The second question investigated by Tiverton concerned the size and location of the striking force. The size of the force, roughly two thousand bombers, was based on projections given to General Smuts by Lord Cowdray drawn from figures presented to the War Cabinet by Sir William Weir, controller of aeronautical equipment in the Ministry of Munitions.[35] This long-range bombing force, Tiverton concluded, should be based in eastern France and assigned to at least two clusters of aerodromes to avoid "dangerous and vulnerable crowding."[36] The third issue he addressed was the need for this massive force to concentrate its efforts on a systematic approach to the destruction of key German targets. Tiverton calculated that a force of 1,630 airplanes could fly 1,086 sorties every other day, delivering 30,400 bombs on German targets every week—a level in his view that would prove sufficient to the task. Lastly, Tiverton concluded that the moral effects of the attacks upon the

German workers could be as significant as the destructive effects of the bombs themselves.

On September 24, 1917, the Gothas returned to begin what are now called the "Harvest Moon" raids, conducting five night attacks on London in eight days. The bombings, which were conducted with impunity, caused the British public to be outraged by the country's seeming defenseless-ness against night raiders. The War Cabinet quickly arranged for defenses around London to be reinforced with anti–aircraft artillery diverted from France, but politicians, the public, and the press clamored for reprisals against German cities.

After discussing the matter with Trenchard, the cabinet on October 2 authorized immediate deployment of a small force to begin active opera-tions against German industry in reprisal for the raids on London.[37] This became No. 41 Wing, formed from two RAF squadrons recalled from other duties and eight Handley Page 0/100 bombers from "A" Squadron, RNAS.[38] It was based at Ochey in France and commanded by Lt. Col. Cyril L. N. Newall. Newall's meager force would not be augmented with additional aircraft until May of the following year. The Wing mounted its first attack against a factory at Saarbrucken on October 17, 1917, with eleven Airco DH-4s. A week later, nine Handley Pages from "A" Squadron carried out a night attack against the same factory while sixteen Royal Aircraft Factory FE-2bs bombed railways nearby. Bad weather limited the Wing's activities and it was only able to execute fourteen missions before the end of 1917.[39]

Strapped for aircraft to undertake the strategic bombing mission, the British looked to the Americans for support, hoping that they would respond enthusiastically to Tiverton's proposals, which he forwarded to Pershing's headquarters via his American counterparts on the fifth floor at 45 Rue Montaigne, Paris.[40] But Pershing and his staff, like their British counterparts in France, were determined to keep aviation carefully coor-dinated with the ground effort. The AEF would not commit to the British plan until they had formulated their own operational concepts.

Major Gorrell was the likely conduit for the information Tiverton passed to the Americans. Gorrell's position as head of the Air Service Tech-nical Section and the close proximity of the two men's offices would have fostered such an exchange. George Williams, an Air Force historian who conducted a detailed study of Tiverton's influence on Gorrell, believes that their personalities were closely matched. Williams' examination of Gorrell's staff work revealed a "gradual convergence between many of his views and those of his British mentor."[41] The exchange of detailed information began in the weeks after the American's return to Paris at the end of July when Tiverton provided him with a twelve-page mathematical treatise titled "Probability in Relation to Bomb Dropping."

By the middle of October, Gorrell had become a dedicated proponent of strategic bombardment. By then the official program for AEF aviation projected sixty bombing squadrons. Gorrell was certain that the U.S. Air Service would wreak immense destruction on enemy morale and material if it managed to field a large number of squadrons to carry out a "systematic bombardment" of Germany.[42] But the one thousand bombers contained in the official building program were not enough in his mind. At least three thousand would be needed to do the job, perhaps as many as six thousand. As I. B. Holley Jr. noted, "Gorrell left little doubt as to the importance he attached to bombardment." Writing to Bolling on October 17, Gorrell concluded that "this [the strategic bombardment of Germany] was not a phantom nor a dream, but is a huge reality capable of being carried out with success if the United States will only carry on a sufficiently large campaign for next year, and manufacture the types of airplanes that lend themselves to the campaign, instead of building pursuit planes already out of date here in Europe." Had Gorrell's recommendation been followed, it would have changed official policy. But it was not. As Holley also pointed out, the program that would determine the composition of the Air Service by functional type of aircraft would undergo "violent changes" between October 1917 and June 1918.[43]

Gorrell's duties as head of the Technical Section of the AEF Air Service ended abruptly in the third week of November after Brig. Gen. Benjamin D. Foulois and a select staff of experts arrived to assume command of the American air force in Europe. Before taking over as chief of the Air Service on November 27, 1917, Foulois worked to prepare an air policy that would provide General Pershing with "a definite policy of operations for the Air Service."[44] Included were plans for a proposed bombing campaign that Gorrell submitted the day after Foulois took command. An analysis of the memorandum authored by Foulois for the Chief of Staff at this time suggests that he knew Gorrell and others were already working on such a proposal.[45] Within a week of its submission, Gorrell was promoted to lieutenant colonel and detailed to Foulois' headquarters as head of Strategical Aviation in the Zone of Advance.[46] His assignment was to establish the staff and provide the planning necessary to field the fifty-five bombing squadrons that the Strategical Section was expected to deploy in the summer of 1918.

Gorrell's proposal to bomb Germany drew heavily upon Tiverton's earlier work, but Tiverton was not the only British expert consulted. One of Gorrell's first acts as head of the Strategical Section was to request the services of Wing Commander Spenser Grey whom Gorrell later claimed was then "the world's greatest authority on questions dealing with aerial bombardment."[47] Grey, formally the commanding officer of No. 5 Wing, was removed from flying duty and detailed by his Royal Navy superiors to assist the U.S. Air Service after he was found unfit to fly as a result of injuries sustained during his aviation career.[48] His assignment with the U.S. airmen began on October 30, 1917.[49] The timing of Grey's appointment coincided with Gorrell's growing interest in strategic bombardment and the Admiralty Air Board's attempts to convince the AEF Air Service to join with British forces in their plans to destroy the German munitions industry by aerial bombardment.

Gorrell's tenure as head of the Strategical Aviation Section proved short-lived, however. His report on strategic bombing had been received favorably by Foulois who initially recommended that the Air Service cooperate with the Allies in the campaign against German industry planned to begin in 1918.[50] But Foulois reversed himself in January after Col. Wilson B. Burtt, the newly appointed assistant chief of staff for Air Service policy, submitted a report on the aerial program needed to support the AEF.[51] At Pershing's insistence, the Air Service, with the Allies' consent, agreed that its first priority should be aircraft for battlefield operation. Burtt concluded that no additional aviation units should be brought over to France until an American Army—including its aviation—was on hand. This derailed Gorrell's plans to assist the British in strategic bombing as the extra squadrons needed for this purpose would "unduly delay the formation of a balanced army."[52]

Unaware of the rapidly shifting sands in U.S. aviation policy, Gorrell submitted his own memorandum a day after Burtt's report appeared.[53] Instead of integrating American air power into Pershing's grand plan for the AEF, Gorrell focused on cooperating with the independent air operations being advocated by the British. This action did not sit well with Foulois, who removed Gorrell from the Strategical Section by reassigning him to the general staff as aviation officer in the G-3 Operations Section.[54]

Although planning for the U.S. bombing program continued after Gorrell's departure from the Strategical Section on February 5, 1918, Spenser Grey recognized it would be some time before any American squadron would be available for the strategic mission. He soon realized that he could render more service to the U.S. Navy, which was then in the process of debating whether it should concentrate its air efforts on "a continuous bombing offensive against enemy [submarine] bases."[55] Grey's contributions to his new friends would exert a deep and long-lasting influence on the Navy's ongoing efforts to crush the hornets' nest.

Chapter Six
The General Board Speaks

January–March 1918

E ven as the Allies and the U.S. Army debated and planned their various
strategic bombing programs, the Navy engaged in a detailed reassess-
ment of its plans and priorities. In fact, by the winter of 1917–18,
a combination of factors caused many to question the feasibility of conducting anti-submarine activities from NAS Dunkirk and seek alternatives.
Part of the concern revolved around difficulties encountered operating the
northern outpost in the face of relentless enemy bombardment and the
seeming ineffectiveness of ongoing search and patrol missions. Offensive,
land-based operations conducted by RNAS units in the region offered an
intriguing alternative.

Coincidentally, the United States Marine Corps' strong desire to find
a role "Over There" for its aviation units then training in various locations
throughout the United States (Mineola, New York; Lake Charles, Louisiana;
Philadelphia, Pennsylvania; and Miami, Florida) added a new factor to the
equation. During the summer the Marines determined to send an infantry
brigade to France, which would require training and dispatch of at least one
reconnaissance and artillery spotting squadron.[1] Toward that end, Marine
aviator Capt. Alfred A. Cunningham undertook a European inspection tour
in the fall and early winter. Upon his return, he submitted a lengthy report
advocating a major shift in the anti-submarine campaign. Equally important, Lt. Kenneth Whiting, leader of the First Aeronautic Detachment and
originator of many proposals underlying the massive, evolving naval aviation program in France, had just reached the United States with orders from

Admiral Benson to assemble men and material to establish a huge seaplane bombing base at Killingholme, England.

Aware of all these developments, the General Board conducted hearings in January and February 1918 on the issue of anti–submarine efforts, taking testimony from Captain Cunningham, Lieutenant Whiting, and Cdr. John Towers who was Capt. Noble Irwin's principal aide in the Office of Naval Aviation. In late February, the Board called for a major new offensive effort to employ hundreds of seaplane bombers and land-based escorts to seize control of the air over the North Sea and hunt and attack submarines departing or returning to their bases in Belgium. Secretary Daniels and CNO Benson, newly receptive to the offensive possibilities offered by naval aviation, responded quickly, issuing orders in early March establishing such a program.

As had been the case since early in the year, the continuing submarine threat ranked foremost in Navy planners' minds. Though the toll of sinkings receded following the spring 1917 crisis—thanks in large part to the institution of the convoy system and improved patrol measures—losses continued at an alarming rate, with no end in sight. Between August 1917 and January 1918 more than 440 vessels were lost to enemy torpedoes, mines, and gunfire.[2] With plans to dispatch millions of troops to Europe in the new year, ending the submarine scourge remained at the center of naval efforts.

Difficulties attendant to placing NAS Dunkirk in operation and the now obvious challenges inherent in the physical and military situation there raised questions of when and if the station might ever be able to carry out its anti-submarine mission. Even as personnel strove mightily to place their station in operation while undergoing perpetual bombardment, Marine Corps aviators back in the United States labored equally energetically to create an aviation force and find a mission for it. Was there an alternative to fruitless overwater patrolling? This question would weigh heavily in the General Board's eventual recommendations.

Alfred Austell Cunningham, the pioneering Marine pilot who had taken the lead in promoting the Corps' wartime aviation program, was born in Atlanta, Georgia, in 1882. He served in the Spanish-American War as a teenage member of the Second Georgia Infantry Volunteers. He later attended Gordon Military College and then pursued a career in real estate. In 1909, he joined the Marines as a second lieutenant, serving on board the battleships *New Jersey* and *North Dakota*, and the receiving ship *Lancaster* before being ordered to Marine Barracks in Philadelphia where he experimented with a primitive aircraft. On May 22, 1912—the date Marine Corps aviation designates as its official birthday—Cunningham reported to the Navy's flying camp at Annapolis, Maryland, for instruction. He soloed later that year after receiving additional training at the Burgess Company factory in Marblehead, Massachusetts. Cunningham's widow Josephine later recalled, "As the Marine Corps' first aviator, he had a one track mind. His life and soul belonged to Marine Aviation." In 1917, he was the de facto director of Marine Corps aviation.[3]

By early 1917, a small nucleus of Marine aviators existed, including Bernard Smith, William McIlvain, Francis "Cocky" Evans, and Roy Geiger.[4] On February 26, Cunningham began organizing the Marine Corps Aeronautic Company, the service's first tactical unit, initially intended to cooperate with the Advance Base Force, utilizing both sea and land planes. The outbreak of war in April 1917 and subsequent decision to dispatch infantry units to France brought with it efforts to expand the Corps' aviation strength and prepare for service overseas. The existing small aviation unit became the Marine Aeronautic Company, tasked with flying anti-submarine missions with the Navy.[5] Maj. Gen. Commandant[6] George Barnett, in an effort to provide aerial support for his ground troops destined for France, authorized formation of a second unit, employing land planes to fly artillery spotting/reconnaissance missions.

Needing to secure up-to-date information regarding training and equipment and to assess evolving military conditions on the Western Front, Barnett in October 1917 ordered now Captain Cunningham on an extended inspection trip of aviation schools, facilities, and operations in France and Britain. Developing an expanded anti-submarine or strategic bombing initiative was definitely not part of his assignment or part of Corps thinking.

The Marine officer departed New York November 3, 1917, on board the SS *St. Paul*; the crossing proved trying, the weather cold and stormy.[7] A gale hit the ship after two days at sea, with waves breaking over the forecastle and solid water "racing down the promenade deck." Thereafter the seas calmed. Shipboard activities included studying French "without much success," a shuffleboard tournament, gunnery practice, being photographed, playing poker (he lost $4.50), changing from uniform into civilian clothes, and looking for submarines while standing two-hour watches. Along the route, *St. Paul* rendezvoused with destroyers *Conyngham* and *Jacob Jones* "camouflaged to the limit."[8] Cunningham called it "a most strenuous five days and am all worn out." He landed at Liverpool November 12 and had lunch at the Adelphi Hotel, complaining about a quarter lump of sugar, no butter, very little bread, and a portion of filet that resembled "a piece of tripe." From Liverpool it was on to London and the Savoy Hotel.

Cunningham spent several days in London, shopping for uniform accessories, writing to his wife, reporting to the American embassy/Navy headquarters, and attending the theater. He also visited the aircraft center at Hendon and Madame Tussaud's gallery, and eventually secured accommodations on one of the cross-channel steamers cruising from Southampton to Le Havre. The Marine flyer arrived in France November 17 and traveled to Paris and the Grand Hotel. After reporting to the American embassy, he checked in at Navy headquarters where he encountered fellow aviators Kenneth Whiting, Frank R. McCrary, Virgil Griffin, and Carl Hull. From there it was on to Army headquarters and then U.S. Air Service offices where he spoke to several officers, trying to arrange permission to visit French flying facilities. After a bit more sightseeing, he set out for Tours on November 23.

Cunningham began taking detailed notes early in his trip. While still in Le Havre, he recorded observations of a French seaplane station and its equipment. At Tours, formerly a large French training site now taken over by the U.S. Army, he counted ten large wooden hangars and several canvas Bessaneau hangars, Cunningham found Avord, another major training site, "larger than all the other schools in France combined," with eight hundred aircraft, twelve flying fields, and three thousand mechanics.[9] He returned to Paris November 28, and after a short stay traveled to Pau. Cunningham observed acrobatic flying, commented on student death rates, and then continued down the coast to the French gunnery school at Cazaux where he had dinner with eight escadrille commanders. Back in Paris once more, he encountered Lt. Godfrey Chevalier, inspected the aircraft depot at Villacoublay, visited AEF headquarters at Chaumont, and met with more Army air personnel.

In mid-December, Cunningham traveled to the front with the French 4th Army, went over the lines in a biplane SPAD, and visited a sector controlled by the French 6th Army. It was during these weeks when he met with U.S. Army officials that he learned they were unwilling to allow a Marine observation squadron anywhere near the fighting. As Cunningham recalled, the Army "stated candidly that if the [Marine] squadron ever got to France it would be used to furnish personnel to run one of their training fields, but that was as near to the front as it would ever get."[10]

Thoroughly disappointed and sick much of the time, Cunningham began a trip to Dunkirk December 24. He toured the American station as well as neighboring British and French seaplane bases. He also inspected the RNAS depot at St. Pol and drove out to Coudekerque to see the Handley Page bombers stationed there. He then hitched a ride to Dover on board a Royal Navy destroyer and traveled to London to meet with Lt. W. Atlee Edwards, Admiral Sims' aide for aviation. After a few days in the city, he journeyed to the RNAS training field at Eastchurch, followed by a visit to a gunnery school at Hythe, where he went aloft, fired at targets, and shot at another plane with a camera gun. Cunningham departed Liverpool aboard the SS *St. Louis* on January 6, 1918.

The voyage home proved far worse than Cunningham's first crossing. Authorities held the vessel in port at Belfast due to the lurking presence of several enemy submarines. Setting out again on January 7, the ship encountered a fierce gale, "with the heaviest snow I ever saw." Wind and waves washed a member of the gun crew overboard, but "we could not even try to pick him up." The ship rolled so badly Cunningham could barely remain in his bunk, even as his trunk "would get loose and slide around the stateroom making an awful racket." During the long crossing, still ill and homesick, he commenced work on a document describing his trip, observations, and recommendations.

Cunningham's lengthy report incorporated information on a wide range of subjects, including an extended discussion of French pilot training, bombing and gunnery instruction in England, descriptions of French and British squadrons at the front, bombing squadrons, Villacoublay, the SPAD factory, carrier pigeons, and the equipment of a French patrol unit. The report's most important elements concentrated on the American station at Dunkirk and conditions affecting aviation in the vicinity, followed by his recommendation for employment of Marine forces in the region.[11]

Cunningham described Dunkirk harbor as "really nothing more than a narrow basin bordered by quays and congested with traffic." It was not, he noted, "a good location for a seaplane base," but must serve as nothing else in the area was available. Conditions at the station were crowded, it lacked sufficient bombproofs, and German aircraft attacked whenever the weather cleared. According to British pilots Cunningham interviewed, the station's patrol bombers should not venture northward along the Belgian coast unless escorted, but Hanriot-Dupont planes acquired for the task were completely inadequate. When the British attempted to operate along the coast the previous fall, their planes "were shot down as fast as they were sent out." The last time the British attempted to seize control of the air had occurred in August. The station lost a three-plane patrol and the rescue boat sent in search of survivors. German land planes had meted out such punishment.

Cunningham believed, however, that naval aviation's greatest opportunity lay in just these contested skies, over the narrows of the English Channel and the waters off the Belgian coast. He noted how the Germans maintained large submarine facilities at Bruges, Zeebrugge, and Ostend, and vessels based there, to reach their hunting grounds, must transit the English Channel. Various shoals and sand bars in the region required U-boats to operate part of the time while surfaced or awash. At those times, they became vulnerable to detection and attack by aircraft. The British, however, with enormous demands placed on their overstretched aviation forces, including protection of London, could not spare the resources necessary to tackle the Germans in the Belgium region. Night bombing raids were neither frequent nor concentrated enough to put submarine facilities "out of business."

In Cunningham's analysis, naval aviation faced several challenges, including seizing control of the air and keeping it. Patrol bombers at Dunkirk could attack submarines in transit, but must be protected by aircraft far more capable than Hanriot-Dupont floatplanes. Cunningham then made several specific recommendations. More space should be secured for operations at Dunkirk, with various facilities, shops, and hangars dispersed as protection against bombing raids. Patrol planes should be used primarily for "offensive work against submarines," depending on fighting machines for protection. Currently, a Marine flying squadron was being organized for land flying. Originally intended to operate alongside the Corps infantry brigade in France, that mission had now been set aside. The force consisted of 31 officer-pilots and 148 mechanics. Cunningham suggested the unit "can be of much more value if used at Dunkirk to assist in securing control of the air and protecting the machines operating against submarines."

In other words, the Marines had flyers ready to fight but no place to go. They should now be prepared to operate at Dunkirk, equipped with a land machine being developed by the British incorporating various flotation devices and watertight compartments. A field within ten miles of the coast should be secured to accommodate the unit. As the work increased in intensity, a second squadron should be organized to assist the first. To train newly purposed anti-submarine units, he recommended that the Marines take over the Curtiss flying school near Miami, not far from the large naval air station, which could provide logistical support.[12]

Coincident with Cunningham's mission to Europe, Kenneth Whiting had been working feverishly to implement a joint Royal Naval Air Service/ United States Navy plan to create a large base on the east coast of England to carry out long-range bombing raids against German fleet targets in the Heligoland-Wilhelmshaven-Cuxhaven area. These efforts would employ specially built lighters, towed by destroyers, and capable of carrying multi-engine flying boats within striking distance of enemy targets. The Admiralty first broached the possibility of a joint British-American campaign in September and October 1917. Captain Cone and Lieutenant Whiting inspected possible sites for an American base on the east coast of England, eventually choosing Killingholme on the banks of the Humber estuary. In November, Admiral Benson approved the plan and issued orders to begin preparations. Whiting returned to the United States to organize the work.

Based on the crucial need to suppress the submarine threat, the slow pace of progress at NAS Dunkirk, Captain Cunningham's recent trip and report, and Navy approval of the Killingholme lighter/bombing initiative, the General Board in January and February 1918 conducted hearings on the subject of "aeronautical stations, and particularly our own people and what they are doing against the submarine menace."[13] The Board had been created on March 13, 1900, pursuant to General Order No. 544, issued by Secretary of the Navy John D. Long to "insure [the] efficient preparation of the fleet in case of war."[14] Led for many years by Admiral of the Navy George Dewey, its members consisted of senior officers charged with offering recommendations on issues ranging from broad strategy to ship characteristics. That the Board evidenced considerable interest in the application of air power to the submarine threat came as little surprise. For several years, the group had examined aviation and the role it might play in fleet activities and naval strategy. Several senior members expressed considerable support for aeronautic endeavors.

Rear Adm. Charles J. Badger and Rear Adm. Robert G. Winterhalter conducted most of the questioning during these hearings. Badger was a native of Maryland and the son of a Navy commodore. He graduated from the Naval Academy in 1873 and served during the Spanish-American War; other duties included a tour as Superintendent of the Academy and command of the battleship *Kansas*. Badger ultimately rose to lead the Atlantic Fleet and later served for several years as chairman of the General Board, retiring in 1921. Winterhalter was born in Detroit and graduated from the the Naval Academy in 1877. An officer of marked scientific bent, he read or spoke eleven languages. At the age of forty, Winterhalter married Broadway star Helen Dauvray. He later commanded the battleship *Louisiana*. Winterhalter served as Aide for Material, the principal assistant to the Secretary of the Navy, in 1914–15 until appointed commander of the Asiatic Fleet, a post he held until 1917. Both officers evidenced a marked interest in aviation developments. Badger became an early supporter based on his observations during fleet exercises in Cuba in 1913 and at Vera Cruz, Mexico, the following year.[15]

Kenneth Whiting, Alfred Cunningham, and John Towers offered the principal testimony. Whiting led off on January 16. He provided a lengthy summary of the origins of the Navy's air stations in France and Great Britain, followed by a description of conditions at NAS Dunkirk and discussion of the current state of training and logistics. He emphasized the continued and destructive attacks by German bombers. Whiting also testified as to the effectiveness of seaplanes as an anti-submarine weapon. Other topics included British night bombing activities, dependence of the Navy program on increased production of Liberty Engines, the necessity of obtaining a geared version of that power plant, and possible use of the Bugatti engine. Admiral Winterhalter then switched the focus of the hearing toward defense of the American coast and Whiting responded by calling for a massive construction program, building air stations not more than 120 miles apart, each equipped with 24 large flying boats, with rest and refueling stations located equidistant between the large bases. Whiting also offered testimony comparing the effectiveness of French and British patrolling schemes, deeming the British far superior.

The Board reconvened February 5 to hear from Captain Cunningham, who repeated much of the information contained in his recently submitted report. Based on his extended inspection trip to Europe and the detailed analysis and plans he had prepared, Cunningham certainly constituted the most important witness to appear before the Board. Admiral Badger opened the session by requesting that Cunningham offer his observations on the recently completed tour of France and England. The Marine aviator turned immediately to the situation at Dunkirk and the opportunities it offered for naval aviation to assist efforts to curb the submarine threat. He described crowded conditions in the port and impediments to efficient and safe operations. He also outlined anti–submarine operations in the English Channel and along the Belgian shore. Cunningham noted that the presence of numerous shoals in the region caused submarines to surface often, unlike the situation along the coast of France where hunting U-boats was "more like looking for a needle in a haystack."[16]

Cunningham's most important testimony focused on his call for the Navy/Marines to seize control of the air over the Channel, permitting lumbering American patrol/bomber flying boats operating out of Dunkirk to seek and attack U-boats transiting local waters on the way to their hunting grounds off the British/Irish shores and the lengthy coast of France. Britain, he noted more than once, had neither personnel nor equipment to carry out this task and was instead concentrating on attacking German U-boat facilities at Bruges, Ostend, and Zeebrugge through bombing via land-based aircraft. "They seem to have decided that they can do more damage with the limited number of machines available if they concentrate on bombing the bases." Cunningham concluded, "The British are apparently not going to attempt to operate with aircraft against submarines in the contested waters on a large enough scale to succeed," hence the opportunity for the United States to fill the void.

Space to accommodate an American initiative might soon become available, as the RNAS had abandoned their Dunkirk harbor station "because they cannot attempt any longer to get control of the air at that place." Operational demands along the Western Front and the need to protect London against enemy air attacks meant sufficient planes and men could not be spared to contest for mastery of the sky. Badger then put the question directly to Cunningham, would 30 bombing flying boats and 150 fighting scouts be sufficient to wrest control of the air from the Germans and carry out offensive patrols? Cunningham responded with an emphatic, "Yes, sir." A discussion of tactics and the necessity of escorting the patrol craft followed, as well as the general difficulty of trying to operate at night.

Asked by Winterhalter when Marine fighting squadrons might be organized, Marine Corps Commandant Barnett testified the process was already under way. Conversely, Cunningham charged the Army was not going to do anything in this area. In fact, he took the opportunity to take a dig at the same institution that had rebuffed Marine efforts to send a squadron to Europe, noting dryly, "The Army Aviation Corps does not seem to take kindly to helping others out." Estimates of the necessary number of pilots and aircraft were offered, up to 250 flyers per year and over 600 aircraft in the same time frame. These planes would operate from airfields located within five or ten miles of Dunkirk.

In sum, Cunningham called strongly and confidently for creation of several land-based squadrons to battle the Germans for control of the air over the Channel and along the coast while mounting heavy offensive, anti-submarine seaplane patrols into previously off-limits U-boat routes to the Belgian coast, with Marine fighters flying protective escort. The work would entail finding and attacking submarines at sea while in transit from their bases to the western reaches of the Channel, filling a void left by British abandonment of similar efforts in favor of land-based bombing raids against German facilities at Bruges-Zeebrugge-Ostend.

On a parallel track, the very next day Cunningham made a formal recommendation to Barnett concerning Marine aviation in which he claimed, "There is needed for operating with the Navy station at Dunkirk all the land pilots the Marine section can furnish for some time." He urged vastly increasing the size of the Marine Corps Flying Reserve, up to 20 percent the size of the naval aviation force. He also called for acquisition of the Curtiss flying field about five miles from Miami, Florida.[17]

Following Whiting and Cunningham's presentations, the Board heard from Cdr. John Towers. He had served two years in London (1914–16) as assistant naval attaché before returning to Washington in late 1916. Since then he had been the moving force behind the development of naval aviation, even when supplanted by Capt. Noble Irwin in the chain of command. In his appearance before the Board, Towers also discussed the submarine situation in the Belgium-Dunkirk vicinity. He described how U-boats passed through various restricted channels before entering the English Channel and then onward to the French coast. To facilitate this movement, the enemy concentrated substantial aerial assets to sweep the skies clear of Allied aircraft. Whenever the British or French tried to change the balance of forces, the Germans brought in sufficient land squadrons to "overwhelm" Allied efforts. According to Towers, "Allied aircraft don't dare go out very far from their base because they have not been able to get sufficient force to combat the Germans." He then endorsed Cunningham's view that dispatching bombing aircraft without fighting escorts "is a useless sacrifice."[18]

With testimony concluded, the Board wasted little time drafting a significant policy document in the form of a "Proposed Letter" in which they analyzed the situation along the Belgian coast and offered a major recommendation designed to control the air in the Dunkirk-Calais region to prevent passage of the Dover Strait by German submarines. This proposal ultimately sparked naval aviation's largest offensive initiative of the war. The Board began by highlighting the critical role played by submarines based in Belgium in carrying out underwater attacks in the Channel and off the French coast. Efforts by the British to close the strait promised only partial success, while sufficient patrol craft were not available. American efforts to build additional destroyers were "not encouraging." Thus it was imperative "to develop to the limit an air offensive against the submarine," arguing that failing to close the strait would severely undercut the effectiveness of the proposed North Sea mine barrage.

Quoting directly from testimony by Whiting, Cunningham, and Towers, the Board acknowledged the size of the enemy's Flanders U-boat fleet, the congested nature of Dunkirk harbor, and the decision by the British not to contest for control of the air but to concentrate their efforts and equipment on bombing the bases directly. The Board endorsed the notion that attacking submarines at sea by aircraft could be done successfully, but only when escorted by fighter aircraft. The senior officers clearly understood the magnitude of the challenge, noting the proximity of large numbers of land-based German aircraft at Bruges, Rouliers, Lille, and elsewhere (in addition to seaplanes at Zeebrugge and Ostend).

The answer, the Board asserted, would be to employ a mixed aerial armada consisting of thirty large bombing seaplanes and at least two hundred fighting scouts to provide protection, the entire activity being "distinctly naval."[19] The Navy and Marine Corps would supply the pilots, the Navy the seaplanes, and the Army the land fighting machines. The Board placed enormous importance on this initiative, "considering these [objectives] are of such importance as to require priority in aircraft assignment of personnel and material to accomplish them."

Having laid out their reasoning and emphasized the importance of the venture, the Board made the following recommendations to put the plan into practice:

Sites to accommodate thirty seaplanes and two hundred land planes should be selected in the Dunkirk-Calais region.

Efforts to obtain material and train men should begin immediately, with the goal of seizing control of the air "in the shortest practical time."

Plans should be initiated to prepare personnel sufficient for thirty bombing planes and sixty-four land planes ready to operate by May.

Obtain necessary aircraft from the Army or independent contract if the Army did not cooperate. Arrange to increase eventually the number of land planes operating in the war zone to two hundred.

Invite cooperation from the French and the British, especially in supplying necessary aircraft. Proceed with the belief that "there cannot be too many bombing and fighting machines supplied quickly."

The Board concluded by declaring firmly that the Navy should give priority in material and personnel to the new initiative, saying it was "of greater importance in the results to be obtained than any other airplane work abroad or at home now projected by the Navy."[20]

General Board Report 820, with minor revisions, passed quickly to Secretary Daniels and CNO Benson for consideration. Working with remarkable speed the Navy's senior leadership took only a week to recast the Board's recommendations as policy. A "highly confidential" March 5, 1918, cable from Benson's office to Navy headquarters in London (received March 7) directed Admiral Sims to prepare aerodromes in France to accommodate four squadrons totaling sixty-four fighter aircraft, to be used to escort thirty seaplane bombing machines, the number of land-based machines to increase eventually to two hundred. Their purpose would be to secure control of the air along the Belgian coast so that German submarine traffic through the Straits of Dover could be harassed and curtailed. The directive also included the complementary mission of interfering with operations from German air bases.[21]

Other cables followed, including another from Benson on March 10, indicating Secretary Daniels had approved the program and training (in the United States) was under way, as were plans to procure necessary equipment.[22] Benson went on to say that aircraft would be secured from the Army, but any obtained from Britain or France would prove invaluable, as he anticipated delays shipping material from the United States. Sims was advised that any aerodrome sites selected should be difficult for enemy bombers to find, in order to avoid the destructive aerial raids that so crippled activities at Dunkirk. According to Benson, the four land squadrons would be ready by June 1.[23]

With the decision to establish a sizable bombing offensive, the highest levels of the Navy, both civil and military, committed themselves to a massive aviation effort designed to end the submarine menace once and for all. The pronouncement, in turn, set off a flurry of activity on both sides of the Atlantic, in one case to modify the directive, and in the other to secure aircraft, men, and training facilities necessary to implement it. Out of this ferment evolved the Northern Bombing Group, a completely land-based strike force directed not at submarines in transit but at U-boat facilities in Belgium.

Chapter Seven
Paris Charts a
Different Course

January–April 1918

E ven as the General Board took testimony and formulated recommendations, naval planners in France and England carefully exam-
ined the challenge of attacking U-boat facilities in Belgium, but
developed a very different response. Participants in this parallel process
included staff wunderkind Lt. (jg) Robert Lovett, spark plug Lt. Edward
O. McDonnell, aviation commander Capt. Hutchinson Cone, and Admiral
Sims' newly established Planning Section based in London. The RNAS'
Spenser Grey played a pivotal role in these deliberations, as did Capt.
Charles Lambe, RNAS, commander of air units in the Dunkirk region.
Lovett and McDonnell's personal investigations included participation in
at least a dozen dangerous bombing raids across the lines. The plan ultimately formulated in Paris and London advocated continuous, massive
day and night attacks against German targets carried out by large, land-
based bombers flying from aerodromes in Flanders, a policy completely
at variance with the initiative just authorized by the Secretary of the Navy
and the Chief of Naval Operations.

The single individual most closely associated with development of Navy
bombing policy in Europe was Robert A. Lovett, a shockingly junior twenty-
two-year-old officer with less than one year of military service, and much of
that of a very informal nature. Yet he quickly emerged as one of headquarters'

most valued members, especially where gathering and analyzing data and formulating recommendations were concerned. Born in Huntsville, Texas, the
son of Robert S. Lovett, president of the Union Pacific Railroad, Bob Lovett
attended the Hill School in Pottstown, Pennsylvania, before entering Yale University in 1914. There he joined with Trubee Davison, Di Gates, and several
others in 1916 to form a group of neophyte flyers that would become the First
Yale Unit—the first section of reserve pilots in the U.S. Navy. Lovett soloed in
August that year at Port Washington, New York.[1]

Following further training with the Yale Unit at Palm Beach, Florida,
and Huntington, New York, Lovett received his designation as a naval aviator on July 31, 1917. He sailed for England two weeks later, one of the first
Navy flyers dispatched to the war zone. Lovett obtained a French army
brevet after passing through an abbreviated course at Tours and was sent
for additional instruction with the French naval air service at Lac Hourtin.
During September, he was assigned to the fledgling U.S. Navy flight school
at Moutchic where his principal duties involved assembling aircraft, serving as chief pilot, and coordinating the arrival/transit of men and supplies.
Within three months he was reassigned to U.S. Naval Air Service headquarters in Paris, where his organizational and analytical skills had already
attracted notice.

In December 1917, Lovett was detached from duty in France and
ordered to the Royal Navy's principal flying station at Felixstowe with orders
to gather as much technical, organizational, and operational information as
possible about the functioning of a large flying boat base. This data would
then be distributed throughout the growing American effort and would also
prepare Lovett to take command of one of the planned U.S. stations on the
coast of France. As the Royal Navy's chief operational aviation station and
site of intense development efforts in the field of large flying boats, Felixstowe seemed the obvious choice for anyone seeking up-to-date information
about anti-submarine tactics.

Lovett approached his new assignment with zeal. A later chronicler of
the Yale Unit described, the "great many ideas sizzling under his cap. . . .
The truth is that Lieutenant Lovett was insatiable in pursuing information
to its lair and a master hand at making use of it." John Vorys, another Yale
Unit member, veteran of two world wars, and an Ohio congressman for two
decades, described how Lovett spent much of his time at Felixstowe "sending big confidential letters to 'Hutch-Eye' Cone and Admiral Sims." Lovett's
original report, eighteen single-spaced, typewritten pages illustrated with

his own sketches, detailed "the whole theory and practice of manning and operating flying-boat stations and the systems of patrol." Lovett included descriptions of the famous "Spider Web" patrol system, handling character-istics of the Felixstowe F2A flying boat, and aircraft armament, ordnance, equipment, and engines.[2]

While conducting research, Lovett received extensive access to reports and statistics compiled by Cdr. John Cyril Porte, RN, one of the world's leading authorities on flying boats, and others at the station. He discovered that Germany had thus far built approximately 350 U-boats, but less than 50 cruised at sea at any given time. The remainder could be found in harbor undergoing repair, rest, and refit. In April 1917 during the height of the submarine assault, only 30–40 U-boats operated in Brit-ish and French waters. Given the limited number of vessels at sea and the difficulty spotting them, each patrolling aircraft covered 22,000 square miles of open waters before sighting a submarine. Only half of sightings led to an aircraft approaching within bombing range. Only one attack in four resulted in a presumed "hit." Lovett credited this analysis to a Brit-ish Captain Cooper, and information supplied to him by the Inter-Allied Command in Paris.[3]

Lovett quickly realized that a strategy based on intercepting U-boats at sea would never end the submarine scourge and the great distance across the treacherous North Sea ruled out sustained attacks on principal bases at Bremen and Wilhelmshaven. Instead, he believed, the German submarine complex at Bruges-Ostend-Zeebrugge on the coast of Belgium, home to the active Flanders flotilla, and where most vessels clustered when not on patrol, was vulnerable to aerial attack. The British had thus far not been able to adequately perform this task. Perhaps the U.S. Navy could.

After completing his stint at Felixstowe, Lovett received orders to attend British bombing and gunnery courses and consult with RNAS Intelligence officers, from whom he gathered additional data and compiled copious notes, all the while strengthening his belief that overwater, anti-submarine patrols were largely a waste of time. Returning to France in mid-January 1918, he assumed a new post as assistant to Capt. Thomas T. Craven, chief of operations in the Navy's Paris headquarters. Lovett's principal duty in those weeks entailed coordinating and facilitating efforts to establish sea-plane stations along the French coast. In the evening, however, while many of his fellow Paris-based officers joined the city's frantic night life, Lovett, by his own words, prepared a plan "for the formation of a group which could be employed in attacking submarines and naval bases, concentration points, naval stores, and factories." Here lay the true origin of the Navy's strategic bombing concept.

Up to this point, Lovett's aviation experience related almost entirely to flying boats and seaplanes, from his days training with the First Yale Unit, to assembling Franco-British aviation flying boats at Moutchic, to inspections of F2As at Felixstowe.[4] While developing ideas about attacking German bases he naturally envisioned using such aircraft, but recognized their limitations. An encounter with Wing Commander Spenser Grey during a January meeting of the Allied Air Council held at the Paris headquarters changed everything.[5]

In fact, Grey was well known to the Navy. As early as August 1917, Lieutenant Whiting and Naval Constructor Westervelt of the Bolling Commission had met with Captain Lambe and Wing Commander Grey during a visit to Dunkirk. Westervelt was so impressed with the British bombing expert that he urged he be recruited to aid the Navy's fledgling aviation program. Writing directly to Admiral Sims, Westervelt suggested, "A very able officer of the RNAS who would be of the greatest assistance on your staff is Commander Spenser Grey RNAS now in command of Naval bombing operations in France." In fact, Grey's eventual contributions proved instrumental in the development of the Northern Bombing Group.[6]

Lovett came to the January meeting in Paris with plans for the U.S. Navy to conduct a seaplane bombing campaign against the U-boat bases in Flanders using the modified Curtiss F-boats developed by Commander Porte, but Grey quickly convinced Lovett that it was impractical to use seaplanes against land bases defended by wheeled aircraft. Land planes, whose speed and performance were not limited by the heavy floats needed for water landings, would simply shoot the attackers out of the sky. To avoid anti-aircraft fire, daylight attacks planned by Lovett would have to be carried out at high altitude, which would reduce the chances of hitting the objective. As Grey had learned during his tenure as head of No. 5 Wing, the solution was night bombing by land-based aircraft.

Grey offered to introduce Lovett to the leaders of the night bombing squadrons at Coudekerque.[7] Grey's arguments convinced the young American to rethink his ideas about the F-boat, which he now realized had limited load-carrying capacity and was designed solely for patrol duty. Lovett still believed that bombing the submarine bases was "the only way to meet and check the U-boat menace," but this should be accomplished using large aircraft such as the Handley Page or Caproni types operated by the British and Italians. As Lovett later explained to Captain Cone, Grey "had done the real work in obtaining the statistics" for a report on bombing that Lovett submitted a few weeks later.[8] Because of the disorganized condition of the Army Air Service, which he characterized as "suffering from too much politics and too little gray matter," Lovett suggested that Grey be attached to the United States Naval Aviation Forces in Foreign Service (USNAFFS), a suggestion that later paid enormous dividends.

Based on these discussions and additional data he gathered, Lovett went right to the top and asked Captain Cone for permission to pursue the proposal, at the cost of foregoing his anticipated command of the planned naval air station at L'Aber Vrach on the northwest coast of Brittany. Instead, he worked feverishly to complete a lengthy memo outlining his vision for destroying the German submarine force in Belgium. Ironically, a conference held at NAS Dunkirk at the same time concluded, "Operations . . . are to be directed definitely against submarines and not against land bases that can be bombed with much greater effect at night with land machines."[9]

Lovett submitted his views in a January 28, 1918, document that ultimately played a central role in the development of the program that led to the establishment of the Northern Bombing Group.[10] He began by acknowledging the size and complexity of the proposal and claimed the entire purpose of the exercise was to spur information gathering and the planning process. He opined that naval aviation could not confine itself to offshore patrol missions and bombardment of U-boat harbors offered the only method capable of ending the submarine menace. Lovett repeated his observation that relatively few enemy raiders actually cruised at sea at any given time and "the wisest way to check them would be to get them in large numbers, which means in their bases." Many years later in a postwar, after-dinner speech, Lovett made the point succinctly when he recalled, "Instead of chasing these Huns separately," it was far better "to try and get them when they were together."[11]

Large seaplanes of the F type, such as those employed at Felixstowe and scheduled for use at the Navy's proposed base at Killingholme, were not designed for either night flying or offensive bombing. They carried small ordnance loads and could operate only during the day, necessitating attacking from great heights. Rather, multiengine, land-based aircraft of the Handley Page or Caproni type should be used, as "the effectiveness of these machines cannot be overestimated." Unfortunately, "No American pilot has ever gathered any personal information regarding them." Since Lovett believed offensive aviation operations would inevitably occupy an important place in the Navy's panoply of strategic plans, he urged in the strongest language that information regarding these large aircraft, both technical and operational, be gathered as soon as possible so they might be incorporated into the service's 1919 program. As Yale Unit chronicler Marc Wortman later stated, Lovett possessed "a young man's brash disdain for tradition and a mind trained to absorb vast quantities of information and deliver practical conclusions."[12]

Lovett went on to make several assertions, what he called "cardinal points." Offensive bombing demanded a specific type of machine, night bombing offered the most effective tactic, and special machines required special pilots trained in both the operation and science of their craft. If these assertions were true, he continued, then "first-hand information must be gathered . . . from an active source," specifically an operational Handley Page squadron, what he called "the most deadly weapon any of the Allies has yet discovered."

Seeing the U.S. Naval Air Service as uniquely positioned to participate in such warfare, Lovett again urged that information be gathered to form the basis for planning future programs. As the author of the proposal, and already the officer most familiar with the theory and reports of night bombing operations, Lovett requested permission to visit RNAS squadrons stationed near Dunkirk to compile additional data, explore aircraft capabilities, and interview pilots. In justifying his possible extended absence from the Paris headquarters, he claimed the current inactive status of most coastal American stations generated few demands on the staff, while "the absence of my slight help could scarcely be noticed." Although it took several weeks, Lovett eventually received orders to depart for Dunkirk to carry out the investigation he proposed in his late January memorandum.

Perhaps not coincidentally, Lovett's Paris memorandum discussing the best way to end the submarine threat dovetailed precisely with a document being prepared by the newly organized Planning Section at Admiral Sims' London headquarters. The concept of creating a staff unit tasked with studying particular problems, criticizing current organization and methods, and preparing plans for future operations had been under discussion for several months. Sims recommended such an idea as early as October 1917. Following his autumn trip to Europe, CNO Benson endorsed a similar arrangement, putting the plan into practice with authorization dated

December 26, 1917. Officers assigned this duty included Capt. Frank H. Schofield, Cdr. Dudley W. Knox, Cdr. Harry E. Yarnell, and Col. Robert H. Dunlap, USMC.[13] In his postwar memoir, Sims spoke very highly of these officers and their contributions to the war effort. All were Naval War College graduates and "scholarly students of naval warfare." Relieved of administrative duties, they focused solely on surveying ongoing operations, identifying errors, and making suggestions for improvements.[14]

The Planning Section got right to work, focusing much of its early energy on examining the ongoing submarine challenge, especially as the prospect of transshipment of millions of doughboys loomed before them. In a February 15 report, the unit identified suppressing the U-boat as aviation's greatest challenge. They outlined three areas where aircraft could be employed: patrolling for submarines, escorting convoys, and attacking enemy bases. Echoing Lovett's assertions that patrol methods had proved generally unsuccessful, they determined, "Our primary air effort should be offensive against enemy bases." This required seizing control of the air from the German air force and then unleashing huge bombers. Seaplane patrols should continue, but only in a supplemental role. These recommendations projected a significant shift away from flying boats toward land planes, and a related movement away from attacks against submarines at sea and toward aerial assaults on permanent land bases and facilities.[15]

Circulated for comment, the report raised several sticky issues, such as competing with U.S. Army aviation programs, possibly scaling back patrol efforts along the French coast, and constructing a vast array of large, new facilities. Captain Cone endorsed night bombing raids, supported initiating photo reconnaissance, and advocated protective aircraft escorts for the lumbering bombers as they flew to and from their targets, while opposing ending anti-submarine patrols. He also noted that seaplane pilots would have to be retrained to secure control of the air and in the necessary bombardment techniques required to attack enemy aerodromes. Cone's comments were surely informed by Lovett's January 28 memorandum, submitted only two weeks before the Planning Section issued its report.[16]

The role of the British in shaping this document seems apparent. According to the postwar history of the RAF, the British Air Council at that very moment intended that a northern bombing group be organized to carry out "large-scale attacks on German naval targets on or near the Belgian coast, notably the U-boat bases at Bruges and Zeebrugge." This group would operate as the counterpart of a planned southern bombing group charged with attacking German industrial centers, what later became Gen. Hugh Trenchard's Independent Force. The northern group never materialized, however, due in large measure to the impact of Germany's spring offensives and British aircraft production shortcomings. Ultimately, most of the technical information available to the American Naval Planning Section were derived from Admiralty sources, as reports describing operational success or failure. The final publication included lengthy remarks from representatives of both the Admiralty and the British Plans Division.[17]

The Planning Section's report, issued well before the General Board's recommendations were made or communicated to Europe, offered additional inspiration for the birth of the Northern Bombing Group. Having just returned from a trip along the French coast, Cone discovered the recently arrived Problem No. 6 on his desk. He described it as "a Peach," noting that "my crowd is busy working on it for next year's plans. It gives us all something to aim at here, which was very badly needed, and I assure you that the business of this organization will be to accomplish as much of this plan as possible."[18]

For a short time there matters sat. Lovett had conducted a series of interviews, carried out extensive research, and submitted a memo asking that the issue of utilizing land-based bombers to attack submarine infrastructure be examined through direct observation. The Planning Section then weighed in, essentially endorsing Lovett's observations. Paris headquarters examined both propositions and made suggestions and counterproposals. Then came the thunderclap from Washington, cables from Benson informing Paris that Secretary Daniels had endorsed the General Board's plan to inaugurate a campaign utilizing large seaplanes, escorted by fighters, to bomb submarines while in transit from their bases to their hunting grounds.

Cone reacted immediately, initiating further research and soliciting reactions from Allied commanders while also sparking extended debate within Navy circles that lasted many weeks and ultimately resulted in abandonment of the Board's seaplane bombing recommendation. Cone dispatched Lovett to Dunkirk to confer with Captain Lambe on the very same day the Department cable arrived "relative to establishing an airplane fighting force . . . to be manned by Marines." Three days later, he ordered Lt. Eddie McDonnell to join Lovett for observation duty with the British.[19]

Twenty-seven-year-old Eddie McDonnell reached the battle lines first. Born in Baltimore in 1891, he graduated from the Naval Academy in 1912 where his determination and pugnaciousness earned him a boxing championship. During the following year, he served on board battleships *New Jersey*, *Florida*, and the armored cruiser *Montana* before receiving torpedo instruction on the destroyer *Montgomery*. During the American occupation of Vera Cruz, Mexico, in April 1914, McDonnell served on board the auxiliary cruiser *Prairie* and for actions while ashore was cited for heroism in battle and awarded the Medal of Honor. In December he began flight instruction at the Wright Company in Ohio and reported to Pensacola on February 1, 1915. He was designated a naval aviator on September 18, 1915. McDonnell remained at Pensacola acting as a flight instructor and check pilot until detached in the spring of 1917 to oversee training of the First Yale Unit in Florida and New York. Completing that duty in August, he assumed command of the aviation detachment at Hampton Roads, Virginia, until ordered overseas. During the winter of 1917–18 he worked at aviation headquarters in Paris, while sharing a small apartment in the Latin Quarter among what Lovett archly described as "the long-haired men and the short-haired women."[20]

Responding to the General Board's proposal, Captain Lambe in Dunkirk claimed it would be futile to try to protect slow flying seaplanes with fast scouts, while highlighting their inability to cooperate tactically during an attack. He argued for bombers that could protect themselves, such as DH-4s flying in formation, what he called "self-defensive squadrons." Lovett returned to Paris March 18.[21] No sooner had Lovett reported to Paris headquarters, however, than Cone ordered him back to Dunkirk, acceding to the junior officer's January 28 suggestion that Navy flyers (specifically himself) be dispatched to front-line aerodromes. When he learned his request had been met Lovett claimed, "My colossal luck took a turn yesterday that has left me gasping in astonishment." He could hardly breathe because he was "so blamed excited. . . . I get cold all night thinking about it, because it's rather scary. Then I get my nerve back and revel in dreams."[22]

Lovett and McDonnell's indoctrination in British practice included obtaining information regarding the future organization of the Navy's bombing squadrons, gaining practical experience operating large night bombers, and locating optimum sites to establish necessary aerodromes. Lovett asserted, "I am going to the front . . . on a wonderful stunt, upon the success of which hangs our future program—the importance of which makes me burst with pride."[23] During the next three weeks, they continued interviewing active-duty personnel at RNAS aerodromes and participated in an extensive series of terrifying bombing raids against German targets in Belgium, the first Navy involvement in such activities. Reports generated by their experiences went a long way toward shaping naval aviation's approach to the issue of using concentrated aerial bombardment by land-based aircraft to eliminate the submarine threat.

McDonnell made his initial cross-lines foray March 20 on board a Handley Page bomber named the "Evening Star" of 7 Squadron, piloted by veteran Flight Lt. John Roy Allan. A total of four, multiengine aircraft participated in the night raid against the docks at Bruges, with McDonnell manning the rear upper and lower machine guns protecting the bomber's tail. After takeoff, the aircraft crossed the coast and climbed, partially masking its approach toward the target. Turning inland and cruising along at 5,800 feet, McDonnell's aircraft dropped nearly 1,700 pounds of ordnance. When one of the bombs stuck in its rack, he shoved it free by hand. After completing its attack, McDonnell's plane re-crossed the target. All aircraft passed through a heavy barrage of shrapnel and high explosives that extended upward to an altitude of ten-housand feet, the machines bucking and jumping from the concussive effect of the blasts. McDonnell also counted sixteen powerful searchlights intent on pinning the slow-moving bombers against the black sky. Later recounting the adventure to his former students at NAS Dunkirk, McDonnell allowed that "night bombing is all right for them as likes it, but that he prefers going walking down home on Sunday afternoons."[24]

McDonnell returned to action on the evening of March 23, rejoining Flight Lieutenant Allan, this time positioned in the forward cockpit acting as gunner. Again they attacked Bruges and again encountered a fierce barrage, being struck three times by shrapnel. Once more one of the 250-lb bombs stuck in its rack. On the return flight, Allan directed his aircraft across the German aerodrome at Aertrycke, Belgium,[25] where the rear gunner shoved the bomb loose when signaled via flashlight.

With very little rest, McDonnell participated in another night raid the following evening. This time as gunner/bomber/observer on board a DH-9 day bomber, Flight Lieutenant Stevens in command. McDonnell's duties included working the high altitude sights, dropping the bombs, and providing defensive machine-gun fire. Six of the underpowered aircraft, in tight formation, each carrying eight small bombs, flew off the Belgian coast as far as Zeebrugge, slowly gaining altitude, then crossed inland at 15,500 feet and headed directly for the Bruges docks. Commencing bombing just before crossing the target, the formation immediately encountered vicious anti-aircraft fire.

McDonnell later recalled, "High explosive was bursting in the midst of our formation with great rapidity and precision. Several machines were hit." One chunk of shrapnel punched a hole through the wing of his aircraft next to the fuselage. Five enemy aircraft jumped the bombers from the rear. In the fight that followed, one German fighter fell before the British guns. Another attacked McDonnell's plane. The American fired numerous bursts at his tormenter, with tracers passing just beneath the fuselage, whereupon the enemy turned away. The three remaining Germans shadowed the departing bombers all the way back to the lines, staying just out of range, hoping to pick off a straggler. All aircraft returned to Coudekerque safely, though one sustained heavy damage.

As the German army poured through shattered British defenses in Flanders, British bombers shifted targets and on the evening of March 26, McDonnell participated in a large night raid by ten Handley Pages (his fourth and last raid) against the vital railroad junction at Valenciennes, about ninety miles south of Dunkirk. The lumbering aircraft cruised at an altitude of 10,000 feet, with visibility very good and the air very cold, and encountered anti-aircraft fire along most of the route. Returning to base, the machine passed over the city of Lille, only to be caught in a blinding searchlight. With his pilot unable to escape the illuminating beam, McDonnell opened fire with the rear lower machine gun, directing tracers down the beam of light. He claimed he shot the light out; his friends and compatriots were not convinced. Bad weather forced his aircraft to descend to 3,000 feet, where they became lost before they found their way to the coast. They eventually located Calais, and from there flew on to Dunkirk. The raid lasted ten hours and two bombers failed to return. McDonnell returned to Paris March 28, then traveled to London to meet with Admiral Sims' staff before being ordered back to Washington at the beginning of April.[26]

On the morning of March 21, only hours after McDonnell returned from his first bombing raid, Lovett departed Paris with a motorcycle and sidecar, reaching Dunkirk after an exhausting, 185-mile, all-day ride. He reported to his girlfriend and future wife Adele Brown, "I am off on a great stunt. The only fly in the ointment is that I cannot share it with you." That evening at dinner, he reunited with his Yale compatriots at NAS Dunkirk. The following day, Lovett joined 7 Squadron, No. 5 Wing, RNAS, a Handley Page unit based at Coudekerque aerodrome. By March 23 he had been assigned to the aircraft commanded by the Canadian Allan. Lovett found the British flyers "a fine lot—the most fearless, clean-cut, happy men imaginable—and awfully good fun." He appreciated Allan's dry sense of humor, calling "good old Roy Allan" an "intimate" friend, adding, "I'd go anywhere with him."[27]

That evening, he participated in a four-plane attack against Bruges, the same mission McDonnell joined. Despite a smoke screen, visibility over the docks proved excellent. All aircraft landed safely, braving the heavy ground mist that descended around midnight. Lovett now found himself caught up in the ferocious rhythm of the bombing assault against the enemy submarine complex, at one point joining four raids in five days, including two daylight attacks. Other sorties targeted non-naval targets. On two occasions, foul weather, rain and hail, forced the bombers to turn back short of their objectives. Altogether, Lovett participated in eight raids in a two-week period. Another mission against the Bruges docks on the night of March 31 proved particularly unsettling. Several of the numerous searchlights guarding the city caught Lovett's aircraft. When the pilot made a second run from the opposite direction, laboring to release a bomb caught in the racks, the searchlights found them again. Defensive machine-gun fire by Lovett caused two of the searchlights to shut down.[28]

Lovett's private description of these missions is revealing. He claimed, "The earth seemed to split, the sky was alive with green onions and high explosives and Archie made a wail through which we dove for our objective." The enemy seemed "to have little difficulty finding the correct height almost immediately upon warning of our approach." After one bomb run he reported, "I had the distinct satisfaction of seeing our bombs lift some ammunition stores sky high." On another occasion, while engaging searchlights with his machine gun, the normally reserved Lovett turned manic and began singing, "Have you seen the ducks go for their morning walk? Quack, quack, quack, quack, quack, quack." Continual firing turned the "tip of the gun red and the glow crept up the barrel." Safely landed, he reported, "I could have cried I felt so lucky to have been through it and come out safely." Kenneth MacLeish offered the observation that Lovett had "the wind up so badly he can't stand still." Shaking with excitement, and downing malt balls to calm his roiling stomach, Lovett told Adele, it was "the greatest experience of my life."[29]

While attached to 7 Squadron, Lovett and McDonnell undoubtedly met and may have flown with Paul Bewsher, a Handley Page pilot, poet, and author who captured these exact experiences in his postwar memoir *Green Balls: The Adventures of a Night Bomber*, or one of his many wartime poems, including

The Dawn Patrol and especially *The Bombing of Bruges*, dedicated to pilot and friend Roy Allan. Relying on the cloak of darkness to carry out his missions, Bewsher dedicated *Green Balls* to "MY FAITHFUL FRIEND, WHO DURING THE WAR PROTECTED ME FROM THE ENEMY AND A THOUSAND TIMES SAVED MY LIFE, THE NIGHT SKY."[30]

In addition to carrying out combat missions, McDonnell and Lovett met with area commanders to discuss organizational issues. On March 22, they again interviewed Captain Lambe. He strongly endorsed the concept of the U.S. Navy joining the bombing offensive, recommending employment of six day and six night squadrons, to be used in a concentrated manner against a single target until destroyed. Each successive target would be attacked in a similar manner. The young American officers spent the following day visiting possible aerodrome sites in the Gravelines vicinity. Captain Cone arrived on March 26 accompanied by McDonnell and Godfrey Chevalier. All three met with Lambe to continue discussions about the Navy's bombing initiative.[31]

Amassing technical documents constituted an important element of this inter-Allied cooperation. Lovett's report on operations included notes on squadron organization, an outline of officers' duties, procedures for carrying out raids, lists of equipment, descriptions of rigging and assembling Handley Page aircraft, data concerning overhaul of Rolls Royce motors, and an overview of control of a Handley Page squadron during a raid. Lovett departed Flanders for Paris April 11. Roy Allan invited him to make one more flight, but the young American could not schedule it. That very night, Allan's plane was hit, caught fire, and crashed into the sea off Ostend. He drowned before rescuers reached the downed machine. Only two of six aircraft returned from the raid.[32]

The experiences, interviews, and investigations of Lovett and McDonnell soon generated a detailed analysis of current British bombing operations and very specific recommendations regarding the size and scope of the proposed Navy campaign. Lovett described the concentrated defenses at Bruges in great detail, including the famous "flaming onions," a series of "luminous green balls attached to one another somewhat in the manner of chain shot," though not terribly effective. Of more concern were shrapnel and high explosive shells easily capable of downing an aircraft. This ordnance was fired into the air in what Lovett called "box-like layers," known as the barrage, a kind of aerial area defense. He noticed, however, that as successive raids occurred the intensity of anti-aircraft fire waned noticeably. Given a few days' respite, however, the enemy replenished ammunition stocks and the rate of fire returned to its lethal maximum.[33]

Raids against rail lines encountered different sorts of defenses. Instead of utilizing barrage tactics, the Germans stationed highly trained crews at railway junctions with orders to shoot directly at attacking aircraft. Lovett reported, "It is surprising how accurate the enemy aircraft fire was. They seem to have little difficulty in finding the correct height almost immediately upon warning of our approach. Their searchlights are far more powerful than those employed by the Belgians and the French."

The limitations of contemporary aeronautic technology further complicated matters. Slow, heavy Handley Page 0/100 bombers, when carrying full ordnance and fuel loads attained speeds of only 75 or 80 mph and proved incapable of climbing above 7,000 or 8,000 feet, requiring silent attacks "from a glide" at a height of only 5,000 or 6,000 feet. This made the large machines easy targets for gunners on the ground, though Lovett observed that only rarely did the engines or crew compartments sustain damage, with the majority of hits to the wings and tail.

The most important portion of Lovett's report, submitted in early April, consisted of a series of recommendations, each supported by his typically tightly reasoned analysis. He called for "Continuous bombing of one objective," a tactic that would lead inevitably to depletion of anti-aircraft ammunition supplies, diminished morale of gun screws, and declining defensive effectiveness. By launching continuous day and night raids, from progressively lower altitudes as defenses became degraded, and by mixing incendiary bombs with high explosive ordnance, Lovett predicted submarine bases would be "made untenable if not stamped off the map."

Lovett warned against intermittent bombing, diverting squadrons away from their intended missions and allowing the enemy to restore his defenses. He also stressed, "I cannot too strongly urge" the need for expertly trained pilots to command night bombers, men proficient in geography navigation, and the tricks of night flying. This would require establishment of a special school in the United States. Geography of the area could be taught by using "large topographical models." A British bombing expert should command this operation.

Well aware that foul weather frequently bedeviled aviation operations in Flanders, Lovett stressed the need for a robust meteorological service, charged with providing information regarding wind velocity at various altitudes, weather conditions along the coast, and a system for rapidly disseminating the most up-to-date data. Dangerous situations generated by bad weather and almost perennial ground mist could be partially ameliorated through creation of secondary/emergency fields equipped with landing lights.[34]

Lieutenant McDonnell's observations paralleled Lovett's. He also called for continuous raids and advocated use of thermite [incendiary] bombs "to entirely destroy the place," predicting that after a series of attacks "anti-aircraft and fire working parties would be demoralized and *hors de combat.*" McDonnell observed that though it was hard to gauge the amount of damage done by Handley Page raids against submarine bases, enormous German investment in an intricate system of anti-aircraft defenses revealed the enemy's fear of the attacks. This reasoning led McDonnell to propose creation of six night squadrons each equipped with ten heavy bombers and six day squadrons with eighteen DeHavilland DH-9 aircraft.

McDonnell went on to recommend that the first Navy squadrons work from British aerodromes, flying with their allies to gain experience and then forming their own operational units. In the meantime, he urged rapid purchase of one hundred Handley Page or Caproni bombers and two hundred DH-9s, this to be followed by opening a bombing school in the United States. Until such a school functioned, however, the best pilots and observers should be assigned to existing Allied Handley Page and Caproni squadrons. Finally, he suggested that aerodromes be selected and commanding officers for the first squadrons chosen so they might begin the process of perfecting their organizations and securing personnel and equipment.[35]

Significantly, both McDonnell and Lovett's reports hewed almost exactly to the specific proposals first offered by Charles Lambe a few weeks earlier, even to the number, size, and type of squadrons necessary to constitute the new bombing force. Their notions of the impact such bombing would have on the performance of German gun crews and firefighters also reflected British assumptions about the effects of aerial assault on enemy morale.

If the General Board, Secretary Daniels, and CNO Benson had expected that their specific plans for a bombing campaign against German submarines would be welcomed on the other side of the Atlantic, they were badly mistaken. Rather, aviators in France and England proposed an entirely different scheme with a different objective requiring a different form of aircraft and training.

Chapter Eight
The Great Debate

March–May 1918

The General Board's seaplane-centered, overwater bombing plan reached Paris in early March where it initiated a fierce debate lasting nearly three months over the ultimate mission, composition, size, tactics, and equipping of the proposed initiative. Participants in this discussion included factions within the Navy Department, including the Marine Corps, both in Washington and Europe, the War Department and its representatives on the ground in France, the Royal Navy, the newly established Royal Air Force (RAF), and the French command. Not until early June did a final resolution of these contentious issues emerge. Originally envisioned totaling twelve squadrons, the group ultimately designated the Northern Bombing Group would consist of eight squadrons of land-based bombers, split equally between day and night formations. They would conduct continuous day and night raids against enemy submarine facilities, a revolutionary Navy program completely unimaginable just a few scant months before.

As early as March 18, Captain Cone informed Navy headquarters in London that "after mature study conducted in the theater of action proposed," including conferences with Allied commanders operating there and with Allied air authorities, the seaplane bombing proposal was "not advisable." Seaplane patrols should not be escorted by Marine flyers. Instead, a force of two hundred land planes, tentatively divided into six night and six day squadrons, should be allocated to the task of continuous bombing of enemy submarine bases at Zeebrugge, Ostend, Bruges harbor, Bruges Canal, and Ghent. A force of chasse machines could provide high offensive overland patrol. To carry out this task, necessary squadrons should be established.[1] A few days later, he responded to Washington's directive that necessary aircraft be secured abroad, observing such equipment would come at the expense of the Army and efforts would be made to obtain them in England, France, and Italy.[2] Admiral Sims forwarded these objections and recommendations to Washington on March 23, identifying continuous day and night bombing of submarine bases as the primary objective, with a secondary effort directed at bombing enemy aviation centers.[3]

Unwilling to abandon the seaplane proposal, and wary of upsetting the Army, the Navy Department debated recommendations coming from Europe for two weeks before replying on April 4.[4] The Department reiterated the importance of closing the Dover Strait to submarines and claimed attacks against U-boats transiting coastal shoals were not being pursued vigorously enough. Plans from Washington envisioned sufficient seaplanes to carry out the task, to be protected by land-based fighters. If such operations could not be mounted from Dunkirk, then seaplane bases should be shifted to England. Aircraft shipped from the United States would be suitable for day bombing and protection. Night bombers would be ordered, but likely not be available until late fall. In the meantime, Paris should undertake to secure thirty-two chasse machines and forty day bombers by July 1, thus keeping the notion of fighter escorts alive.[5]

Further discussion and consultation carried out by naval aviation officials in France only strengthened Paris' opposition to the General Board's seaplane bomber proposal. Cone met with Col. Richard C. M. Pink, RAF, director of marine operations, who advised against creating a large seaplane force. "I think a great part of this personnel could be better employed," he told Cone. Pink also considered Dunkirk to be "a most uncomfortable place to operate from," and that "a big organization of the type suggested will be extremely lucky if it is not bombed to bits." A better choice would be to create a large force of light bombers of the DH type as "these at least could look after themselves."[6]

Shortly after receiving Lieutenant McDonnell's report in late March, Admiral Sims ordered the energetic lieutenant back to Washington to expedite the flow of material, aircraft, and personnel from the United States to Europe. At the same time, Naval Air Service headquarters in Paris digested Lovett and McDonnell's recommendations, conducted further inquiries, and on April 17 Cone forwarded to Sims a detailed list of requirements for the proposed bombing group "for the establishment of forces to successfully bomb submarine bases." He requested 6 squadrons of heavy night bombers (90 Handley Pages or Capronis), along with 180 officers and 1,764 enlisted men, to include 72 pilots, 72 observers, and 144 gunners. Planners estimated wastage at 20 aircraft per month. To carry out day missions, Cone recommended 6 squadrons equipped with 18 DH-4s or Bristol F.2b type aircraft (total 198—108 in the field, 36 at repair base, 54 at Pauillac), along with 282 officers and 1,266 enlisted personnel, including 120 pilots and 120 observers. A bomber repair base should be established behind the lines, requiring 40 officers, 1,960 men, and 120,000 square feet of industrial floor space in nine large structures and a variety of smaller buildings.

Such an enormous undertaking required a vast array of supplies and logistic support, including tents to accommodate 2,000 men, hardware, stoves, and roofing for 5,097 men and 527 officers, medical supplies for 5,000 men, scores of staff cars, trucks, tankers, motorcycles, ambulances, truck-mounted searchlights, lighting material for all shops, hangars, and barracks, sufficient generators to power the lights and other equipment, fire extinguishers, and 7,000 large gasoline drums. Communications would be provided through creation of telephone networks requiring switchboards and 250 miles of telephone line. Lengthy lists of ordnance requirements, tools, and supplies would follow shortly.[7]

This recommendation was forwarded to London where Lt. W. Atlee Edwards, Sims' aide for aviation, recast the material, which was then cabled to Washington, along with the admiral's remarkable endorsement that "inasmuch as bombing operations are considered to be the primary mission of the U.S. Naval Aviation Forces, Foreign Service, strongly recommend approval of this entire project."[8] At the same time, Cone requested Department authorization to contract for Caproni aircraft in Italy, sufficient to equip night bombing squadrons.[9]

Finally, after much additional discussion, Secretary Daniels capitulated and approved the Northern Bombing Group on April 30 in the form advocated by Paris headquarters. The General Board's original plan would be "extended" to include night bombing land machines and additional day bombing aircraft; the Army had been consulted and agreed such activities were "purely naval work."[10] Operations required 6 squadrons of night bombers, 6 squadrons of day bombers, including 240 day bombers, 120 night bombers, 90 Curtiss JN-4 training aircraft, and 10 Thomas Morse advanced training machines. Navy personnel would operate 6 night and 2 day squadrons, while Marine Corps aviators would man 4 day squadrons. All bureaus were ordered to expedite shipment of personnel and material.[11]

As CNO Benson explained, "These operations are considered the most important of any yet undertaken by Naval Aviation," and should be given precedence over other activities when necessary. He directed the Navy technical bureaus and offices to expedite shipments of personnel and material. That Daniels, Benson, and Sims identified the bombing initiative as naval aviation's highest priority speaks volumes about the change in attitude at the highest levels of the Department. Marine aviator Cunningham, who advocated so strongly for offensive aerial missions in the Dunkirk region during his February testimony to the General Board, sensed the Navy meant what it said. He informed Capt. Roy S. Geiger, then training flyers in Florida, "They have taken up this work in earnest here and have organized a Department here to secure the necessary supplies."[12]

While planners and commanders on both sides of the Atlantic debated the size, composition, and mission of the proposed bombing effort, an internal struggle raged in Washington, pitting ambitious Navy and Marine flyers against each other. According to Cunningham, when he reached headquarters in April 1918, he discovered that McDonnell had just arrived from France, carrying "an elaborate scheme for the Navy to take over land flying, and take it in such a way as to eventually put us out." The acerbic Cunningham sniped, the coastal seaplane bases (in France) were "more or less washouts" and now the sailors wanted to co-opt his idea for land-based missions! A fierce internecine struggle ensued. The acknowledged though unofficial head of Marine aviation believed that if he had reached Washington just one day later "things would have been in an awful mess."

Both Cunningham and Lt. Douglas Roben, then serving in Washington, strenuously opposed the Navy's supposed poaching of the Marine mission, but made little headway. Increasing the pressure, they enlisted Major General Commandant Barnett, USMC, and his Chief of Staff, Col. Charles G. Long, in a fight that lasted until the evening of April 29. At this final meeting, the Marines "brought things to a showdown." In the end, the Navy approved plans to train four Marine squadrons for day bombing work and employ them in Flanders. The leathernecks also received promises of more supplies and aircraft. Cunningham concluded, "I was extremely worried for several days, but feel now that everything is entirely satisfactory." A formal agreement executed May 2 allocated four day bombing squadrons to the Marines, reserving two others, as well as six night bombing squadrons for Navy aviators.

Marine formations would train in the United States; Navy units would be organized in Europe.[13]

Secretary Daniels' authorization of the new program and the division of labor agreed upon in Washington did not, however, end discussions. Suggestions that Northern Bombing Group activities might include attacks on enemy aerodromes or other expanded operations unsettled both the Navy Department and the Army Air Service. Sensitive to charges the Navy was poaching Army equipment and missions, Daniels on May 28 fired off an explicit directive to Sims. He charged that it appeared plans were being formulated in Europe to greatly expand the use of day and night bombers well beyond the approved mission.[14] Nothing but enemy naval activities were to be targeted. Any aircraft procured abroad must be obtained in cooperation with the Army, and only in numbers to meet short-term needs.

Even more dispiriting from Paris' point of view, Daniels and Benson acceded partially to Army objections (in view of "the vital necessity of equipping Army squadrons on the Western Front") and slashed the proposed size of the Northern Bombing Group by 33 percent, eliminating two night and two day squadrons. Navy training would also be limited only to personnel necessary for the reduced complement, to be replaced by Marines when conditions permitted. Daniels closed with a warning: "Be governed accordingly in plans for future naval aviation activities." In other words, maintain the smaller force structure, stick to the prescribed assault on submarine infrastructure, avoid other land targets, and hand the job over to the Marines as soon as possible. Giving credence to the idea that much of the opposition originated with the Army, Secretary Daniels sent a copy of his dispatch to the War Department. Sims forwarded the same missive to Cone three days later.[15]

Many associated with the bombing project tried hard to reverse the decision to reduce the size of the Northern Bombing Group. Lieutenant McDonnell had been laboring in Washington in the Office of Naval Aviation and traveling around the country since April, attempting to gather up men and equipment for the venture. He was well aware of discussions under way to prune the program, and after oral arguments to the contrary proved unsuccessful, responded on June 2 with a lengthy, vigorous defense in a memo addressed directly to CNO Benson. McDonnell came right to the point. Forthrightly stating his credentials to speak on the subject and his awareness of the challenges faced in implementing the bombing proposal, he observed, "I feel it my duty to urge in writing that the plans as outlined by Adm. Sims and brought to this country by myself be not modified, but be continued as originally proved by the Navy Department."

The feisty lieutenant forcefully outlined justification for the program and the universal support it enjoyed among the various air forces in Europe, especially after extensive consultation with the Allies. He included detailed commentary from Capt. Charles Lambe, as well as descriptions of his own experiences carrying out bombing missions. McDonnell believed that "supplied with the proper number of planes and equipment, bombing each of these objectives [German U-boat facilities] day and night, it would be quite possible to destroy completely each objective in turn."

McDonnell also reacted to pressure exerted by the Army to restrict Navy access to aircraft, "respectfully" inviting attention to the fact that the 90 heavy bombers and 198 day bombers requested formed "a very minute part of the Army's program for these types of planes." In fact, Army plans for the fall called for the delivery of thousands of day bombers and would begin to take delivery on an order of 1,000 large night bombers. He concluded by asserting, "Compared with these figures, the small numbers required for this important work for the Navy cannot be considered a very large order or anything that would embarrass the delivery of planes to the Army."[16]

Captain Cone joined the debate as well, strongly denying promoting any mission creep and asserting the entire project incorporated Army and Allied cooperation. He claimed a force of twelve squadrons was necessary to carry out the mission of assaulting submarine bases. Aircraft bombers contracted in Italy could be directed to the Army if necessary. No reduction in training should take place because pilots and observers could be used by the Allies in an emergency and Cone was proceeding with preparation of flying fields near Dunkirk-Calais sufficient to accommodate twelve squadrons.[17]

Admiral Sims made two final attempts on June 8 and June 12 to reverse the decision pruning back NBG plans, cabling the Navy Department and writing directly to Secretary Daniels. He flatly denied any efforts were under way to expand the bombing program beyond parameters already outlined in detail and approved by the Department. Aware that much turmoil had been stirred up by the Army, he asserted the "entire project" had been developed "after thorough cooperation and consultation with the U.S. Army." Sims took great pains to assure Daniels that contemplated activities were "entirely separate from those contemplated by U.S. Army Aviation." Offering innocence as a shield against Army charges, the admiral averred, "The mere fact that we as sailor-men will operate land machines . . . does not necessarily indicate that we have in anyway encroached upon the Army's territory. If land machines appear to be more desirable in the successful operation of naval aviation there is no reason why naval aviation should not employ this type of machine." As a sop to the Army, however, he suggested if the Navy's bombers were actually needed for duty on the Western Front, they could be easily diverted to such missions.

The size of the proposed group—six night and six day squadrons—had been determined after thorough investigation on the ground and lengthy consultation with British forces in the region. Clearly challenging the Department's directive to pare the size of the force, Sims stated, "It is still believed that the conclusions reached were sound and six of each of these squadrons will actually be necessary for the successful accomplishment of the mission that we have undertaken." He concluded by recommending training of naval aviators for land aircraft "should continue as energetically as possible."[18]

Reflecting the continuing cross-purposes at which Washington and European headquarters sometimes worked, Director of Naval Aviation Irwin advised Admiral Benson that while Sims' letter was not completely in accord with "their apparent intentions from previous recommendations," it nonetheless showed "they [Sims and Cone] have our point of view now." The Department remained unmoved by Sims' missives, no reply was forthcoming, and the cuts proceeded as mandated. Marine aviator Cunningham observed, "As usual plans are being changed so often that it is hard to keep up with them . . . no one knows how long this change will be in effect before another one is made."[19]

One other issue underlay the reluctance of some elements of the Navy Department to proceed with the program: the issue of pilot instruction. Should naval aviators actually be trained to operate land planes? Heretofore the service had preserved a single-minded commitment to seaplanes. Yet information from Europe, especially from Britain, indicated much work currently assigned to seaplanes could be done equally well with land-based aircraft. Later that summer Cone endorsed that view in a personal letter to Irwin, remarking, "I am of the opinion that it will not be very long before we will restrict very greatly the use of seaplanes and use land planes for many things for which we use seaplanes at this time."[20]

Such an approach raised two thorny issues: should the Navy employ land-based aircraft, and if so, how would they cooperate with the fleet? Of more immediate concern, if the Navy operated land planes from land bases, would that not provide even more powerful ammunition to those championing amalgamation of Army and Navy air services as had just occurred in Britain with the creation of the Royal Air Force? A review conducted within the Department of the Navy Northern Bombing Group situation conducted at the end of the summer stated baldly, "That brings us to consideration of the question of amalgamation. The Navy does not want it; Whether the Army does or not is uncertain. . . . If, later, they decide to make capital of our operating against the [submarine] bases to assist amalgamation, the thing is already done."[21]

The sticking point had always been relations with the Army. Up until the General Board's February recommendation, the Navy had virtually no involvement in land flying, save for a small group of enlisted personnel who received primary training at Allied schools in the opening months of the war. And even they soon transitioned to seaplane instruction. Nor did the Navy concern itself with attacks against land targets. This had always been considered Army work. But plans to organize a large strategic force composed of day bombers, night bombers, fighter escorts, and training machines, multiple aerodromes and supply/repair facilities, raised the specter of intense inter-service competition for aircraft, supplies, and missions. It also edged into the acrimonious subject of amalgamation, something to be avoided at all costs. As historian Clifford Lord observed, "One reason for the Navy's hesitancy at taking the new departure was the fear that it might furnish an argument for the amalgamation of the Navy with the Army."[22]

As early as March 7, 1918, Secretary Daniels requested the War Department provide a small supply of land-based aircraft to begin implementing the General Board's recommended program. Included were thirty primary trainers, ten advanced trainers, and eighty Liberty-powered chasse machines. Several months earlier on November 8, 1917, the Aircraft Production Board had declared, "All air measures taken against submarines should have precedence over all other air measures," a policy endorsed by the Secretary of War on November 14.[23] Daniels now noted, "This material is urgently needed to carry out an offensive program against enemy submarines during the favorable weather of the coming spring and summer." Secretary of War Newton Baker replied quickly that necessary material would be turned over to Navy inspectors at the factories as soon as available. On March 12, 1918, the chief of the Army Aircraft Division informed Irwin that delivery of thirty JN-4 trainers and ten Thomas Morse scout/trainers could begin immediately and eighty chasse aircraft sometime later.[24]

On April 4, Daniels officially informed Secretary of War Newton D. Baker of the General Board's February 25 recommendations, explaining the plan to seize control of the air in the Dunkirk region so that heavier bombers could be deployed against submarines transiting the Straits of Dover and shallow waters off the Belgian coast. Daniels also reported on changes recommended by officials in Europe and advanced the service's belief that use of land-based night bombers to attack submarines in their bases constituted purely naval work. He therefore asked Baker, did the War Department agree? If it did not agree, was the Army prepared to undertake this mission themselves? And if the War Department agreed that the mission was, indeed, naval work, was it prepared to supply seventy-five night bombers and forty chasse machines by July 1, and seventy-five two-place fighter/day bombers by October 1? Daniels requested an early decision.[25]

Responding promptly, Baker returned a letter six days later agreeing with Daniels' position, though leaving the question of Navy bombers attacking submarine bases as opposed to attacking submarines in their bases somewhat vague. In reality, the actual practice of aerial bombardment, especially at night, made this a distinction without a practical difference. The Department informed Sims of the exchange three days later on April 13.[26] The Army responded to appeals for additional aircraft by saying it could not supply the first two requests by July 15, but could deliver the fighters by October 1. Finally, they suggested Navy officials in Europe contact General Pershing regarding the need for night bombers and chasse machines.[27]

Further refinement of Navy plans, including acceptance of the notion day bombers could protect themselves and seaplanes would not be used for bombing missions, led the Department on April 27 to cancel its request for chasse aircraft. Extended negotiations regarding delivery schedules for previously sought aircraft followed. Irwin spoke directly to Maj. Gen. William L. Kenly, soon to be named director of Military Aeronautics. On May 2, Secretary Baker suggested a compromise concerning the eighty aircraft promised in March and an additional seventy-five requested in April. Four machines would be turned over immediately for Marine training purposes. Two weeks later, the Navy asked for ninety additional Jennys (Curtiss JN-4s) in lots of twenty and thirty, deliveries to be completed by August 15. The Army responded that this would seriously impact their own training program, but would deliver one-sixth of all production until they satisfied the Navy's request.[28]

All these endeavors had been directed toward acquiring sufficient aircraft to instruct and equip Marine Corps aviators assigned the responsibility of carrying out planned daylight anti-submarine infrastructure bombing raids by Daniels and Benson back in early March. Ultimately, efforts to obtain training aircraft proved modestly successful and comparatively amicable, but access to training facilities remained more problematic. During the fall and winter, the Army accommodated Marine pilots at Hazelhurst Field on Long Island and Gerstner Field in Louisiana, though some leathernecks complained they did most of the work themselves, being largely ignored by their hosts. Forty Navy crews took the gunnery course offered at the RFC-run school at Camp Taliaferro (Hicks Field) in Fort Worth, Texas. When in the spring of 1918 the Marines sought places for four of their officers and twenty enlisted men at an Air Service school in Dayton, however, the Army refused, citing a shortage of equipment. In fact, on June 5 General Kenly and Col. Henry Arnold, assistant director of Military Aeronautics, informed Director of Naval Aviation Irwin that no training fields could be turned over to the Marines, nor could any leathernecks be instructed at Air Service facilities.[29]

Determining the ultimate mission and composition of the Navy's proposed bombing program proved easier said than done and it took more than three months for the process to run its course. Strong differences of opinion within the Department, and between advocates located in Washington and Europe, had to be resolved. Competing tasks, such as anti-submarine patrolling and convoy escort assignments, demanded attention and resources. So did efforts under way in Ireland, England, Italy, western France, and along the coast of the United States. Aspirations of Marine Corps aviation played a large role, while the views of the Allies needed to be considered. The attitude of the War Department, General Pershing, and the U.S. Air Service loomed large in Navy deliberations. Fears of amalgamation and charges of "mission poaching" weighed heavily on planners' minds. Difficult personnel and logistical issues needed to be resolved, as did the question of aircraft procurement, specifically what types of war planes, and from what sources. Resolution of this final challenge would determine the ultimate success or failure of the entire venture.

Chapter Nine
Night Bombers Needed

*We lacked absolutely [any] knowledge of aeronautical military require-
ments. In fact we had not built a single land combat plane of any description either for ourselves or the Allies. We were as ignorant as a child unborn
of the nature of the equipment of a military "plane."*

—Theodore Knappen, *Wings of War*

Once the mission and organization scheme for the Northern Bombing Group had been settled, those in charge of implementing the
plan turned their attention to acquiring the aircraft needed to
equip the twelve (later reduced to eight) squadrons allocated to the project. On this endeavor would hinge the success or failure of the entire program. Aircraft of all types were in short supply, but the situation for night
bombers was particularly critical. Only two examples—the Italian-built
Caproni and the British-built Handley Page—carried a bomb load large
enough to inflict the level of damage envisioned by the Navy. Neither
airframe manufacturer had been able to produce the numbers needed
to fill Allied air objectives and had turned to the United States for help.
Orders for both types had already been placed, but technical problems
and contractual difficulties delayed production well beyond the original
delivery schedules forcing the Navy to approach the Italians, bringing the
sea service into direct competition with the U.S. Air Service over the issue
of Caproni deliveries.

As the RNAS discovered during their abortive attempt to strike at the German heartland in the fall of 1917, large numbers of aircraft were needed if significant damage were to be achieved against a particular target—both on an individual mission basis and for the duration of an air campaign. Large formations of bomb-laden aircraft were needed in order to saturate an area with enough ordnance to devastate a target. To prevent the enemy from repairing damage to such installations it was necessary to conduct repeat missions against the same target over a significant time period. Continuing missions of this nature drastically increased the level of attrition requiring large numbers of new aircraft to replace those that had been destroyed, damaged, or worn out. Thus the need to procure many night bombers remained the critical aspect of proceeding with the Navy's plan.

When the Bolling Mission issued recommendations for the procurement of various airplanes types at the end of July 1917, it selected the Caproni bomber for the long-range night mission. Although nearly everyone was optimistic about the Americans' ability to quickly fill the production gap for Allied aircraft, Bolling—from his observations of aircraft manufacturing practices in Europe—felt strongly that many pitfalls and troubles lay ahead. It was "absolutely necessary" to procure aircraft from Europe if the United States were to participate in the air campaign planned for spring 1918.[1] He urged that the United States purchase all the aircraft the Allies could turn out to tide over American forces until July 1918, the earliest he believed deliveries of aircraft from the United States could commence. It would be much wiser, Bolling felt, if the United States furnished the raw materials, which were much easier to ship than finished products.[2]

At Pershing's request, Bolling took charge of purchasing airplanes and engines from the French and Italians until production from America arrived in Europe.[3] One of his first actions was to negotiate a verbal agreement with the chief of Italian Aviation for two hundred three-engine Caproni biplanes with the new Isotta Fraschini engines. This arrangement was never finalized, however, and no aircraft were delivered under that agreement.[4] The Caproni company had tried unsuccessfully to license its triplane bomber to the United States government a few months earlier, but had been rebuffed due to the huge fee demanded by the manufacturer.[5] Although no agreement was reached on the interchange of patent rights and cross licensing, the owner of the company, Gianni Caproni, agreed to send one of its triplane machines to America so that its performance and construction could be reviewed.

While Bolling negotiated with the Italians in Europe, Maj. Raffaele Perfetti, head of the Special Italian Military Commission for Aeronautics in the United States, busily lobbied aviation officials for the construction of Capronis in the United States.[6] "A great American squadron of aeroplanes," Perfetti proclaimed, "would decide the war in favor of the Allies in a short time."[7] By then, Congress had appropriated 64 million dollars to carry out the aircraft program proposed by the Joint Army-Navy Technical Aircraft Board.

The board, which was established shortly after the United States entered the war for the purpose of standardizing the designs and specifications of American aircraft, received new direction and greater impetus when the Ribot Cable reached Washington on May 24, 1917. Within days of its arrival, the Board formulated and recommended an aircraft building program be undertaken in the United States based on the numbers outlined in the French premier's proposal. Unfortunately the cable, as received, failed to include any information regarding the proportion of functional types of aircraft to be provided and was misinterpreted due to errors in transmission that occurred between the time the message was drafted and when it was received by the Joint Army-Navy Technical Aircraft Board.[8] In the absence of a doctrinal precept from France, the Board used the 4,500 aircraft and the factor of 2,000 per month for replacement as a purely quantitative guide. As Air Force historian I. B. Holley noted in his seminal book *Ideas and Weapons*, "They resolved the question of composition according to their own ill-defined ideas of doctrine, which tended to attach greater importance to observation and proportionally less importance to bombing."[9] The production plan called for U.S. manufacturers to produce 12,000 airplanes (including replacements and reserves) during the first half of 1918 for service in Europe.[10] Unfortunately, the Board's misinterpretation of the French proposal had far-reaching effects, and only 1,333 bombers were included in the initial program.

The first indication that bombers were needed in much greater numbers than had been recommended by the Joint Army-Navy Technical Aircraft Board did not become apparent until mid-August, when Pershing cabled the War Department to advise that his staff was discussing U.S. production of Caproni night bombers with the Italians.[11] A second cable recommending the need to supply the AEF with four thousand day bombers and six thousand night bombers followed shortly.[12] Germany, according to information in the cable, was preparing to bomb the Allies on a huge scale in the spring. "We must strike first and hard," wired Pershing. Thousands of night bombers would be needed if the ideal number recommended by our allies of fifteen bombers per kilometer of the front was to be met. Both England and Italy were urging the mass production of bombers in America.

Further information on the importance of bombing was provided when Army and Navy members of the Bolling Mission submitted their own report on September 4, 1917.[13] "It is our settled conviction," the report indicated, "that the importance of bombing operations with direct military ends in view cannot be exaggerated."[14] Lt. Col. Virginius E. Clark, an aeronautical engineer assigned to the Bolling Mission, provided his own assessment of bombing's value to the war effort in a separate memorandum addressed to the Chief Signal Officer of the Army on September 12. "I believe that the employment of these machines [night bombers] in large numbers," he wrote "would in a shorter period of time than is possible by any other means, end the war."[15]

At the time, the United States Army Air Service was still part of the Signal Corps under the command of Brig. Gen. George O. Squier, who was in overall charge of procuring aircraft. On September 14, he cabled Pershing advising that two examples of Caproni bombers (one triplane and one biplane) already sent to America were being assembled at Langley Field (they would be flown to Washington in a demonstration flight on September 22), and that the question of manufacturing rights had been settled with the Italian government. Clark, Squier reported, recommended that the United States supply "fittings and all parts of Caproni triplane and Handley Page biplanes for assembly in Italy and Ireland respectively," along with the 12-cylinder Liberty engines then in the early planning stages.[16] Clark wished to start work on both aircraft immediately and asked that Pershing's staff obtain drawings so that construction could begin. The cable requested that an initial order be negotiated for 1,000 sets of each plane. In addition, 250 more of each type would be assembled in the United States.

The Handley Page 0/100 was Great Britain's answer to Italy's Caproni bombers.[17] Like Giovanni "Gianni" Caproni, Frederick Handley Page had attempted to license production of his model 0/100 bomber in the United States shortly after it entered World War I on the Allies' side. William H. Workman, the company's representative in America, came to the United States to promote construction of Handley Page bombers and began negotiations with the White Motor car company to manufacture the planes.[18] On May 31, 1917, he submitted a proposal to deliver all blueprints, drawings, sample materials, and parts to the U.S. Government for the sum of $250,000 and a royalty of 1 percent of the cost for each machine.[19] The Wilson administration decided, however, that it would not conduct any negotiations with individual manufacturers but only with the Allied governments themselves. Nonetheless, Workman continued to push for construction of 10,000 bombers in the United States. His efforts were further hampered by a continuing stream of cables from Bolling recommending against the Handley Page in favor of the Caproni.

Bolling continued to recommend the Italian bomber throughout the rest of the summer and fall. On September 27, he cabled that he did not understand the reason for constructing two types of night bombers, but that he had telegraphed the Italians to release the Caproni drawings, and had cabled the British Air Board to release their drawings and to send a sample of the Handley Page that Washington had requested.[20] In response to the latter, Bolling's representative in England replied that it would be impossible to send a Handley Page machine as all night bombing machines were urgently needed at the front. As for the drawings, Bolling's representative stated that Brig. Gen. John D. Cormack of the British Air Mission in America had been authorized to immediately release the plans.

On October 9, Bolling advised Washington to "concentrate your efforts on Caproni," adding that he considered the proposal to assemble Handley Pages in Europe impractical.[21] Ignoring Bolling's counsel, the Aircraft Production Board—formed in April 1917 to oversee the production of military aircraft—proceeded with plans to build major components of both aircraft in the United States for assembly overseas. The protracted negotiations and technical difficulties encountered in trying to manufacture both aircraft in a timely manner proved more daunting than anyone in the U.S. industrial establishment had foreseen. Neither program proceeded as fast as was initially planned, forcing the Navy to seek other sources for the aircraft needed to fulfill the mission of the Northern Bombing Group.

As events would show, the highly optimistic projections concerning American industry's ability to quickly produce large, complex, European-designed bombers proved to be illusory. The failure to produce the much needed bombers would have major implications for operational planning in France and elsewhere. Nowhere was this more true than with ill-fated efforts to obtain Caproni aircraft. Having already worked with the Italians for several months, Bolling met with Colonel Perfetti at the Hotel Mirabeau in Paris during the evening of October 9 to discuss the construction of Caproni machines and the American factories that were to produce aircraft assemblies to be shipped to Europe. They met again on October 18, but could not reach any agreement until Gianni Caproni arrived in Paris five days later to confer with Bolling.[22]

At the end of the meeting, Bolling, Perfetti, and Caproni agreed that the Americans would manufacture "sets" of bombers in the United States for assembly in Europe. Because of the shortage of labor throughout the war zone, the Italians had hoped that the Americans would be able to run the assembly plants. But as Bolling explained, the Americans had come to Europe to fight, not to work. Caproni would have to find the workmen needed to assemble the bombers. After visiting several potential sites for an assembly facility in France, Caproni returned to Italy to seek his government's help in obtaining the needed laborers.

In the meantime, the Aircraft Production Board, which had heretofore recommended construction of five hundred Caproni bombers, continued to press the Italians through Bolling to provide necessary drawings and technical data necessary to initiate the manufacturing process. The lack of success in this endeavor forced the Board to recommend that the order for five hundred Capronis issued to the Curtiss Company be reduced to fifty. Bolling responded by cable on November 20, advising they withhold consideration of the Italian aircraft until they obtained complete drawings of the plane designed for the Liberty engine.[23]

When Gianni Caproni returned to Paris in late November, he was not happy to learn that for want of adequate drawings the United States was going ahead with the Handley Page instead of his aircraft. Within days he (or his representative) promised Bolling that they would send an engineer to America within a week with detailed drawings of a biplane bomber designed for the Liberty engine. The firm, Bolling was advised, had already completed three airframes, one of which would be shipped immediately, along with a complete set of fittings.[24]

As Col. Sidney D. Waldon, the former automotive executive who now served as a member assigned to the Aircraft Board,[25] later explained:

Gianni Caproni did everything he could to assist us in planning for the assembly of Caproni planes in France, making factory lay outs, giving us data on materials and parts lists, sequence of operations, etc., and finally working out a plan for using Italian labor to be brought to France from Italy and the United States, and under the guidance of Caproni experts, to assemble American made planes. This plan for assembling was to have carried a payment per plane to Caproni to reimburse him for his organization and to some extent, for his expenses in the U.S.[26]

The Americans did not accept this idea. The Wilson administration remained reluctant to pay royalties to foreign manufacturers for war materials manufactured in the United States and decided against assembly in Europe. Even after Caproni returned to Italy on December 12, negotiations continued with Colonel Perfettti and other representatives of the firm.[27] Despite Bolling's admonition that, "It was exceedingly difficult to get any Italian to make a clean cut proposition," Waldon persisted in his efforts to work out a formal agreement with the Italians put down in "black and white."[28]

When Caproni rejoined the negotiations in Paris in the early part of January 1918, he was accompanied by Giuseppi Grassi, chief of the Italian Air Mission in France. Caproni, upset with the decision to cancel assembly in France, argued that the Americans had breached their agreement. This matter was thrashed out at a conference held in Bolling's office attended by Capt. Fiorello H. LaGuardia, a Congressman from New York, then serving in the AEF Air Service in Italy and a friend of Caproni's who had been invited as a referee. Waldon, also present, later recorded what transpired.

> Colonel Bolling and I went over our side of the case before Captain LaGuardia, which Gianni Caproni, and I think Major Perfetti, presented the case for Caproni. At the conclusion, LaGuardia told Caproni that he did not consider Caproni had any grounds whatever for considering that there had been any agreement reached regarding the use of Italian labor in France or covering the plans Caproni had made in Italy on the assumption that he was going to erect Caproni machines in France for the United States Government.
>
> Colonel Bolling, who liked Gianni Caproni very much, told him in the kindest spirit that in going ahead in Italy as he did with his plans, to hire a factory organization, hire refugees and train them, make plans for buildings, etc., etc., just as though an agreement

had been reached, whereas the basis of an agreement only had been discussed, that he, Caproni, had acted like a simple child and without business judgment.[29]

Caproni, still seeking financial gain from his airplane design along with compensation for expenses incurred, proposed a royalty payment of $350,000 that would allow the United States to construct his model 600–900-hp biplane for the duration of the war. In addition, he asked that his company be reimbursed $150,000 to pay for the expenses it had incurred to date and for the three months of technical support from the team of engineers from the Caproni factory—known as the D'Annunzio Mission—that was en route to the United States to assist American aircraft manufacturers. It was headed by Captain Hugo D'Annunzio, a reserve officer in the Italian Army who had formerly been chief engineer of the Caproni plant in Milan.[30]

Bolling made it perfectly clear that he had negotiated an agreement with the previous Italian Minister of Aviation for the free interchange of rights between governments. As Waldon relates, Signore Grassi, in an effort to help Caproni get some of his money back, was inclined to repudiate the agreement of the former minister, whereupon Colonel Bolling stated that if such were the case he would recommend to the U.S. government that it cease further negotiations with Italy with respect to aviation. At a later conference, Signore Grassi, after communicating with his superiors in Italy, stated that the Air Minister would stand by his predecessor's agreement.

At this point in the negotiations, Bolling had reestablished the principle of free interchange of rights and obligations and had agreed for the United States to pay Caproni for the expenses his company had incurred when it sent three of its aircraft to the United States in September. As for the D'Annunzio Mission, it would be handled on the same basis as similar missions being sent by Spad, Gnome, Handley Page, and other companies, despite the exorbitant rates suggested by Caproni, which the United States refused to pay.

Before leaving Paris to return to the United States, Waldon, sympathetic to Caproni's financial needs, told him that although the United States was under no obligation to pay him for his rights, he (Waldon) would present Caproni's case to the Aircraft Board and would use his influence to persuade the Board to extend the same privilege as had been extended to several of the French manufacturers, that is, the payment of $100,000 for the rights to manufacture Italian bombers in the United States. Caproni demurred a little at the amount, which was much below what he had expected, but said he would help all he could whether paid anything or not. Waldon would later write that Gianni Caproni "knew the power of America and his heart was on winning the war."[31]

Upon his return to Washington, Waldon laid the whole matter before the Aircraft Production Board recommending that:

1. Arrangements be made with Capt. Hugo d'Annunzio and his experts for their wages and expenses on the same basis as with other missions.
2. That sample aircraft and material brought over by Capt. d'Annunzio be purchased at the going rates for such material.
3. That Caproni be paid the sum of $100,000 for the rights to manufacture the Caproni machine.

Col. Robert L. Montgomery, a prominent industrialist in civilian life, objected to item 3, recommending instead that $1,000,000 be paid, but that it would be for the two planes, spare parts, and engines, already sent to the United States, together with the expenses of Caproni's 1917 mission. A subcommittee of the Aircraft Production Board concurred and a resolution to this effect passed on February 12, 1918.

When Captain d'Annunzio arrived in the United States on January 17, 1918, with nineteen of the best men from the Italian factory, he was under of the impression that all of the groundwork had been laid so that production of the Caproni bomber could begin.[32] This was not the case. The final design had yet to be determined; the exact number of aircraft to be ordered was still undecided; and no one knew what service would be supplied. Nevertheless, the Aircraft Production Board still believed that domestic manufacturers could supply one hundred "sets" of aircraft a month beginning in July 1918.[33] Although the Curtiss Aeroplane and Motor Company had been selected to produce the plane, an order was given to the Standard Aircraft Company of Elizabeth, New Jersey, to translate the Italian drawings and to make four examples of the American Caproni under Captain d'Annunzio's supervision as to engineering and design.[34]

Work on the hand-built experimental models began on January 25, 1918, when d'Annunzio and his crew arrived at the plant. Difficulties converting the Italian drawings—which relied on the metric system—into the English system used in the United States, along with variations in the standard manufacturing practices used in American aircraft factories versus those employed in Italy, created numerous technical problems that seriously delayed construction of the first experimental model, which did not take to the air until July 4, 1918. In the meantime, the Aircraft Board decided to order 250 of the Caproni Ca 5 type, designed for the Liberty engine. This action did not occur until mid-April when a verbal order was issued to the Fisher Body Corporation, later increased to 500 machines. Almost two more months passed, however, before the first production contract was officially awarded to the Curtiss Company on June 7, 1918, with an order for 500 planes. A day later Fisher Body secured a similar order. The technical and political delays that hindered award of a formal contract to produce Caproni bombers eliminated the possibility of their contribution to the war effort, as neither company could gear up for production before Armistice was declared in November.[35]

Although Bolling continued to insist that the Caproni was the better airplane, aviation officials in the United States initiated efforts to commence production of the Handley Page bomber based on the drawings provided by General John D. Cormack. The project received the first of several setbacks, however, when London reported in the first week of October that the drawings supplied by Cormack were a preliminary set and correct plans would be mailed to America shortly.[36] These were not received until mid-November, at which point the various firms making Handley Page parts were forced to stop work as the new drawings contained many changes. The updated drawings were forwarded to the manufacturers as soon as copies could be made. Work on these was almost completed when a third set arrived on December 30, along with a team of engineers from the Handley Page factory that had previously been requested by the Aircraft Board.[37] After studying the new plans, the manufacturers were disconcerted to discover that a large percentage of the tools and sample fittings prepared according to the second set of plans were useless, forcing them to start all over again.[38]

Meanwhile, negotiations proceeded in England for assembly of American-manufactured parts there on an enormous scale. On December 16, Pershing cabled that the English were endeavoring to arrange construction of 150 Handley Page bombers per month from parts made in the United States.[39] They would furnish construction facilities along with necessary workers estimated at 26,000, but AEF would have to supply the labor required to build the 15 aerodromes required. The latter would necessitate the use of 10,000 construction workers for three months.

Back in the United States, efforts to build Handley Pages continued unabated, with the decision made to construct the first machine at McCook Field under the direct supervision of the U.S. Air Service Production Engineering Department. To expedite matters, it was later decided to assemble the first bomber at the Standard Aircraft Corporation, which had been given a contract to manufacture five hundred of the planes. At about this time, a meeting took place in London regarding the technical questions concerning the assembly and erection of the American-built machines. On January 10, 1918, orders were issued for the balance of the parts and four days later General Cormack asked that six sets of streamlined wire, wheels, tires, and data on propellers be sent from Britain. Although sources for streamlined wire were being developed in the United States, they were not considered entirely dependable.[40]

All the propeller data Britain had, he was told, was related to the Rolls Royce engine and would be useless for the Liberty-powered version being produced in the United States. In order to ensure timely production of the big bomber in the United States, Cormack recommended that at least six experts focus on propeller design at once, as it was very difficult to develop a good one on the first trial. He advised the manufacturer to send the streamlined wire, wheels, and tires immediately; but these were not shipped until April 29.

In the meantime, General Foulois' staff in Europe had completed an agreement with the Air Ministry to assemble enough Handley Page bombers in Britain to equip thirty Air Service bombing squadrons.[41] The United States would furnish Handley Page parts, except linen, which would be deducted from the part of the supply already allocated, increase the shipment of dope for the wings by 20,000 gallons monthly, provide 3,000 laborers to build the five aerodromes necessary, man the three aircraft acceptance parks that were to be built by the Royal Flying Corps, provide three training depots, and pay £670,000 for construction and £50,000 weekly for assembly. The British in addition to supplying linen and the three acceptance parks, were to convert cotton mills and weaving sheds into assembly shops and provide much of the labor and all the technical data.[42]

The AEF agreement with the Air Ministry had little effect on the Aircraft Board's decision to construct and assemble Handley Page bombers in the United States, despite concern over the duplication of effort needed to produce both bombers. Howard Coffin and General Squier raised this issue at a special meeting of the Aircraft Board held February 8, 1918, both of whom objected to production of two types of night bombers. As Colonel Waldon explained (he had just returned from Europe): "Circumstances have been such that we were practically forced to put the Handley Page into production. We tried last summer to get hold of the drawings of the Caproni and to build that over here, but we were unable to do it and the British came through with the offer of their drawings in advance of the Caproni, so that we put that into production."[43]

Production of the British night bomber in the United States received another setback in early March 1918 when the chief engineer sent over to assist the Americans announced that the fittings for the first prototype scheduled for shipment from England in January had not arrived, and it would be necessary to have all these items manufactured in the United States.

Undaunted by this further delay and eager to proceed with the task of producing large quantities of night bombers, the Aircraft Board recommended that a contract be given to the Grand Rapids Airplane Company (an organization of fifteen furniture manufacturers) to provide enough sets of wood parts for one thousand Handley Page machines, five hundred of which were released for immediate production. This action was taken on March 14. Five days later, on March 19, the Board recommended that an order be placed with the Standard Aircraft Corporation for the assembling and packing for shipment of five hundred machines, 10 percent to be fully assembled for flying and testing in this country, the rest to determine the interchangeability of parts. After a month of negotiation with the Air Service Production Engineering Department, Standard Aircraft Corporation received absolute and unqualified responsibility for all parts and completion of the work.[44]

Because of the various technical delays encountered in establishing a design for the Handley Page suitable for construction in the United States, the first set of parts for assembly was not shipped overseas until July 25, 1918. As a result, the optimistic schedule established by Foulois for the spring offensive could not be met and the United States proved incapable of putting any night bombing squadrons at the front before the war ended. All work stopped abruptly the day after the Armistice was signed. No Handley Page squadrons ever reached France, although two such units had completed their training in England.[45]

These formidable production roadblocks in the United States directly impacted Northern Bombing Group efforts in Europe, igniting fierce competition between the Army and Navy for access to the limited output of Italian and British factories. More than any other factor, the failure to resolve logistical impediments would negate nearly heroic efforts to assemble, train, and field the Navy's first strategic bombing force.

Chapter Ten
Putting the Plan
into Motion

May–July 1918

E ven as manufacturers in the United States struggled to produce
heavy bombers and military leaders in Paris, London, and Washington debated the final parameters of the Navy's new strategic initiative, the work of putting the program in motion commenced. Moving the
northern bombing project from proposal to operational reality constituted
an enormous undertaking that proceeded through many phases extending
across widely scattered geographic locales. Much of the responsibility fell on
naval aviation's headquarters staff occupying offices at 4 Place d'Iena (Hôtel
d'Iena) in central Paris. By this time, the organization, headed by Captain
Hutchinson Cone, had reached substantial proportions, including approximately three dozen officers and nearly sixty enlisted men.[1]

Over a period of four months, this group oversaw a process intended
to achieve several objectives, including creation of a permanent command
structure for what would eventually become the Northern Bombing Group.
This required marshaling manpower, acquiring aircraft and equipment,
cooperating with allies and Army partners, establishing aerodromes and
repair facilities, and training personnel. Some activities proceeded more
quickly and easily than others, but by early August substantial forces had
reached the field, construction was well under way, and missions were being
carried out, largely in partnership with the RAF.

Dispatching Lieutenant McDonnell, just returned from his duties at Dunkirk, back to the United States, constituted one of the first specific actions undertaken to implement the bombing initiative. Shortly after reporting to headquarters in Paris at the end of March, McDonnell received orders to head for London where he would receive definitive instructions. Operating at the specific behest of both Cone and Sims, he soon sailed for the United States, charged with organizing and expediting the flow of men and material necessary to put the program into action. He also carried a series of allowance lists with him, identifying the numbers and types of men and equipment needed in Europe. Sims contacted Secretary Daniels directly, indicating the lieutenant was acting as Cone's personal representative and would provide a full explanation when he arrived.[2]

To bolster McDonnell's leverage in Washington, his brief was broad and the authority requested for him considerable. A letter from Cone to Irwin explained the lieutenant "has actually been engaged in offensive bombing expeditions . . . and is fully in touch with the efforts our allies are making in these projects, particularly the work of the Handley Page squadrons." Based on his experience and the responsibility vested in him by Cone and Sims, "It is therefore requested that the recommendations of Lieutenant McDonnell be followed in all particulars, and that full trust be placed in his general knowledge of the situation and in the various proposals he will make." Finally, Cone requested the young aviator "be given authority to advise the various bureaus concerned . . . and that his advice . . . be given weight in the same manner as if it were sent direct from Commander Aviation Forces, Foreign Service . . . that this officer be considered as having full authority to make any arrangements concerning the above mentioned matters." A few weeks later, after McDonnell had settled in, Cone informed Irwin, "My desires are that you keep him as long as he will be of use to you; in fact, I fully realize you will have him over there for some time to fully benefit by the experience that he gained over here." Cone did offer one reservation, noting that McDonnell "was not here long enough to get thoroughly acquainted, but as you well know, due to the shortage of this class of men, I thought it wise to start him home immediately."[3]

McDonnell set up shop in the offices of Director of Naval Aviation Irwin, but made frequent visits to facilities around the country, including a trip to the Wright factory in Dayton to inspect DH-4 aircraft being built for use by Navy-Marine day bombing squadrons.[4] He also supplied Paris with information regarding expected shipment of aircraft from the United States to France and attended weekly conferences held in Irwin's offices designed to coordinate activities of the Department's semiautonomous bureaus. After much hard work, the lieutenant returned to Paris in late July where he began the thankless task of ferrying the Navy's first Caproni bombers across the Alps from Italy to northern France.[5]

Clashing personalities and competing aviation ambitions often caused friction between McDonnell and Marine flyers with whom he worked. In April, his advocacy of Navy day bombing squadrons helped ignite a controversy that eventually included Commandant Barnett and Captain Irwin. Employed as a roving inspector to evaluate the progress of Marine squadrons training in Florida, McDonnell prepared a somewhat negative assessment, noting deficiencies in gunnery and bomb dropping. Lt. Harvey Mims, aviation officer at Marine headquarters who replaced Captain Cunningham in that slot, exploded. "To hear McDonnell talk," he ranted, "one would think he is a little tin Jesus and that the Navy has had every detailed training pertaining to fixed gun work. . . . He is the same little pill he used to be at Pensacola, that is, in my estimation. However, he and I always get along nicely and I always keep my feelings covered while around him." Other controversies revolved around the respective roles to be played by Marine and Navy aviators in the overall bombing program.[6]

Until the nascent bombing group possessed a separate and robust command structure, Paris headquarters carried most of the institutional burden. Even after the unit became operational, Cone's office continued to play a major role overseeing construction, securing aircraft and equipment, and negotiating with British, French, Italian, and Washington interests, as well as acting as arbiter between competing factions within naval aviation itself and serving as a buffer with the U.S. Army. In April, Cdr. Benjamin Briscoe of the headquarters Assembly and Repair Section, aided by Spenser Grey—now in the RAF—developed lists of the aircraft and personnel that would be needed based on the experience of existing RAF DeHavilland and Handley Page squadrons that included allowances for wastage and repairs of aircraft.[7] They also provided lists of the other equipment and tools required, along with plans for buildings prepared by the Public Works Section.[8]

These lists, carried back to the United States by McDonnell and Briscoe, formed the basis for necessary equipment and staffing requests, and thus the material underpinnings of the entire venture.[9] A few weeks later, Lovett in Operations, cooperating with officers in Planning and Intelligence, reviewed the lists with an eye toward purchasing as much equipment as possible in

Europe, especially certain instruments, field equipment, special types of lighting, and signaling devices. The Paris office also coordinated often frustrating efforts to purchase Caproni bombers in Italy.[10]

When necessary, Cone and members of his liaison group served as diplomats, answering questions and smoothing feathers ruffled by the Navy's singular entry into the land bombing business. In May, he held discussions with Maj. Gen. Mason M. Patrick, newly appointed chief of the Air Service, to soothe raw nerves over an Army-Navy conflict then swirling in Italy concerning the conduct of Capt. Fiorello La Guardia and the acquisition of Caproni bombers. In late June, Cone and liaison officer Ens. George Fearing visited General Pershing at his headquarters at Chaumont. Also in attendance was Brig. Gen. James W. McAndrew, Pershing's Chief of Staff. Cone explained the Navy's plans, responded to any objections, and came away pleased that the AEF commander "showed great interest [in the Navy aviation program] and expressed his desire to assist in every way possible."[11]

Also in May, Cone received news that complaints regarding bombing plans were circulating in London and northern France. A communication from Edwards in London claimed Spenser Grey had reported "some misapprehension existing as regards the British Air Force authorities not having been notified of our projects in France for which the Northern Bombing Squadrons have been organized."[12] Separately, General Lambe in Dunkirk suggested Cone formally notify the Admiralty and Air Ministry of the Americans' plans, with Sims to carry the news.[13]

In an effort to keep their allies informed of the progress and scope of the bombing initiative and allay any fears the program might be diverting resources needed elsewhere, Sims on June 12, 1918, contacted the secretary of the Admiralty, informing him that U.S. Naval Aviation Forces, Foreign Service, had established eight day and night bombing squadrons in the Dunkirk-Calais area for the purpose of "destroying enemy naval bases on the Belgian coast." Sims proposed that the units be placed under the command of the vice admiral at Dover, "as in reality they will form part of the Dover barrage."[14]

Exasperated by the entire affair, Cone complained, "There seems to be a great deal of lost motion as between the Air people in London and those British Air people in the Field in the Northern part of France." Cone visited British headquarters the following day, accompanied by U.S. Army liaison officer Brig. Gen. William W. Harts, to confer with Gen. John Salmond, commander of the RAF in the field. He carried with him a memorandum prepared especially for the occasion describing the scope and nature of the bombing initiative. Cone intended to take several copies to "pass out to anyone in the northern part of France who looks like he had any authority about anything." After the talk, Cone reported that the plans with reference to the bombing program were fully explained.[15]

In the spring of 1918, Paris headquarters assumed the structure it would maintain until split between London and Brest later in September. Captain Cone headed the organization as commander, United States Naval Aviation Forces, Foreign Service. Three principal divisions handled staff work: Intelligence and Planning, Operations, and Administration, with a smaller unit attending to medical activities. The Intelligence and Planning Division under the aegis of Cdr. Henry Dinger oversaw issues related to seaplanes, dirigibles, ordnance and instruments, motors, radios, weather, and liaison. The Operations Division headed by Capt. Thomas Craven contained sections devoted to schools, personnel, the Northern Bombing Group, and all operating stations from Ireland to Italy. The Administration Division led by Lt. (jg) Harry Guggenheim exercised authority over supply, public works (construction), assembly and repair activities, and the headquarters secretarial function. All of these units played a role in moving the NBG project forward. An Executive Committee consisting of Cone and his division heads coordinated the work.[16]

Simply choosing a name for the bombing initiative generated considerable discussion. In May, the Executive Committee settled on Northern Bombing Squadrons, with the proposed repair facility designated the Northern Repair Base, and the aerodromes to be called Naval Flying Field A, B, and so forth. Squadrons would be labeled Night Bombing Squadron 1, 2, 3, up to 6, followed by Day Bombing Squadron 7, 8, up to 12.[17] Alternatively, for many months official paperwork referred to activities and locations in Flanders–Pas de Calais comprising the bombing program as the Northern Bombing Region. By August, the program had been renamed the Northern Bombing Group, with a repair and assembly base at Eastleigh known as Base B.[18]

Creating an independent command structure for the bombing initiative demanded considerable attention and drew on a wide range of personnel resources. At first, work was done "in house," within the existing headquarters organization. This included the efforts of Robert Lovett and others in the Operations Division, several civil engineers from the Intelligence and Planning Division, and Harry Guggenheim, head of the Administration Division, as well as the personal attention of Cone and Craven.

Young Lovett carried much of the load. A friend and schoolmate claimed, "He has more ability than any man in the outfit. Without him we would never get anywhere . . . even his commanding officer [Craven] asks him if it's all right to go out to lunch, because on one occasion he went north without telling Bob and Captain Cone gave him an awful calling down." Lovett had a well-deserved reputation for analyzing situations and moving the paperwork along. "And a good example of his ability is this," his admirer noted, "There was a stack of papers about two feet high, they were all reports. It took men who in civil life were getting about $40,000 a year two days to come to any conclusion on the reports. It took Bob just two and a half hours, and every one of the older men agreed with him on every single conclusion but one."[19]

Others besides Lovett's Yale companions commended his work. In fact, Captain Cone offered extraordinary praise. He identified Lovett as "the medium for getting together all matter of personnel, operation, and organization." In early May, he told Irwin, "I have another youngster, Lieutenant Lovett, a Reservist, who is just as good as McDonnell; in fact, he is one of the best officers I have ever seen. He has spent some time up with the Handley Page bombing squadrons and is now out in the field with a Caproni squadron of night bombers that are up here from Italy helping out at this time."[20]

It soon became obvious, however, that even with Lovett's superior abilities, advancing the bombing project required an overall commander and designated support staff. In early May, Cone called for a quick decision on naming a commanding officer for the Northern Bombing Squadrons and emphasized the need for training to commence. "The matter was discussed at some length [in the Executive Committee] but no decision was reached."[21] Two weeks later, Cone named Guggenheim temporary commander, charging him with business management at headquarters so that he became responsible for "proper coordination of the work of various sections and divisions interested in this project." Just three days later, however, Capt. David C. Hanrahan arrived in Paris, having recently been named to lead the Northern Bombing Squadrons. The very next day, he and Lovett received wide-ranging travel orders in connection with this duty. On May 20, Hanrahan, Lovett, and Spenser Grey set out on an extended inspection tour of the region with plans to meet many of the key players.[22]

Called the Iron Duke by some, Hanrahan had no previous experience with aviation. He was, instead, an experienced destroyer skipper who had also captained the Q-ship *Santee*. A tall, mustachioed man of many facets, various observers described him as magnetic, overbearing, both popular and unpopular, wild, jolly, mild, bad-tempered, a tower of strength, a good organizer, a stern disciplinarian, and possessed of a sparkling Irish sense of humor. Following the war, he served as naval attaché in Poland, commanded the light cruiser *Omaha*, and retired with the rank of captain.[23]

With Hanrahan's arrival, a command structure began taking shape. Guggenheim relinquished his duties May 29. A May 31 headquarters directive declared northern bombing business would now be handled the same way as that of the various air stations scattered across Europe; that Hanrahan and his staff would, for the time being, be working in Paris; and they should be consulted and kept informed as to all matters.[24] The pace of activity accelerated in June when Hanrahan returned to Paris from his lengthy tour. Staff officers, flight personnel, ground crews, and working parties were assembled and dispatched to selected sites.

Lovett left Operations, assigned permanently to the Northern Bombing Squadron, and eventually becoming commander of the Night Bombing Wing. Lt. Godfrey Chevalier departed NAS Dunkirk to temporarily assume the role of group operations officer. Assistant Paymaster Stockhausen accompanied Chevalier, filling the position of supply officer. As such, he assumed responsibility for provisioning the men in the field and furnishing material and supplies for camp construction and preparation of aerodromes. Lt. (jg) Hubert Burnham became Public Works Officer. Lt. Cecil "Mike" Murray moved up from the American training facility at NAS Moutchic to serve as armaments officer. An entire retinue of paymasters and clerks joined them, as did medical officers and a dental surgeon.[25] An early July organizational plan prescribed a headquarters complement of sixteen officers and one hundred enlisted men. Hanrahan, Lovett, and the rest of the staff relocated to St. Inglevert near Calais, but returned frequently to Paris to conduct business and attend conferences. Headquarters shifted to Autingues later in the summer. Though Hanrahan experienced some delays filling various positions, by September 1 the process was complete.[26]

What Paris had wrought did not please everyone, most conspicuously Maj. Alfred Cunningham who commanded the Day Wing of the Northern Bombing Group and reached France at the end of July. Though he masked many of his feelings from Navy partners, Cunningham bared his thoughts to fellow Marines back at Washington headquarters. He was distressed that no one in France seemed to know he and his men were coming. There had been no one to greet them or arrange for lodging and transportation. Promised aircraft and equipment had not arrived. Conditions at the fields were primitive and disorganized.

Relatively quickly, Cunningham became one of Hanrahan's greatest detractors and of the Navy's aviation effort generally, though his earliest impressions had proved favorable. He claimed Hanrahan "has so far been very nice." While the NBG commander currently labored in Paris, Cunningham expected him to arrive in the field permanently by mid-August.[27] In early September, however, after several frustrating weeks trying to get his squadrons organized and flying, the Marine's attitude changed dramatically. In a lengthy missive to Assistant Commandant Long, he offered a damning analysis of the Navy's efforts thus far.

Though Cunningham considered Hanrahan a good officer, he "knows nothing whatever about aviation and he has chosen his staff and commanding officers from people who know almost as little about aviation as he does." The Marine conceded the Navy possessed many good aviation officers, including a few in Europe, "but for some reason Captain Hanrahan does not appear to want them in his organization. . . . This is very unfortunate and will undoubtedly make the work here only a partial success as a whole." The Navy, it seemed to Cunningham, was "doing things in an amateurish way."

Looking inside Hanrahan's office, he claimed, "There is a Colonel Grey of the British service who appears to be actually running things for Captain Hanrahan and it has been suggested that he does not wish experienced officers in the organization as this would make his services unnecessary."[28] Ten days later, Cunningham repeated his charges, saying, "They have placed officers on Captain Hanrahan's staff that know nothing about aviation and are so busy trying to learn the rudiments of their work that the work of getting ready to operate as units is almost at a standstill. . . . The whole organization, outside of the First Marine Aviation Force, is in a chaotic state." Cunningham's sharp criticism persisted right until the end of the war. The acerbic Marine seemed to bear no personal ill will toward Hanrahan, however, noting on September 4 that the captain "had shown a disposition to be perfectly fair with the Marine organization. . . . He has certainly not discriminated against us so far."[29]

While many grew frustrated at the seeming slow pace of progress, at least one aviator expressed amazement at the rapid changes under way. Lt. (jg) Kenneth MacLeish, a member of the First Yale Unit, trained at Moutchic, Gosport, Turnberry, and Ayr, flew anti-submarine patrols at NAS Dunkirk, served with 213 Squadron in April 1918, and then received instruction at Clermont-Ferrand. Selected as a potential flight leader in the Navy's initial day bombing unit, he was among the first pilots placed with 218 Squadron in July where he participated in several attacks against Bruges and Zeebrugge.

Just returned to Paris from the front lines, he enthused about ongoing efforts to put the Northern Bombing Group into operation. "It is perfectly fascinating," he reported, "to see things grow down here at headquarters. What was just an idle dream yesterday is a budding reality today, and that's how things go on. You can't imagine the satisfaction of seeing such things. It is impossible to be discouraged. The news is good from every quarter."[30]

Throughout development and implementation of the Navy's bombing plan, British forces played a central role, advising, teaching, supplying, and cooperating, supporting an initiative that mirrored their own strategic vision.[31] The RNAS, and later the Royal Navy faction within the new RAF such as Charles Lambe and Spenser Grey, had long advocated use of "air power" to neutralize the U-boat threat, but were unable to implement a continuous, directed program of aerial bombardment. By bringing the Americans over to their way of thinking and assisting development of exactly just such a campaign, they could achieve their objectives using the U.S. Navy as their weapon.

Officials like Lambe, Grey, and J. C. Porte delivered voluminous files and reports to Lovett and others, helping shape their views of conditions. They engaged in extended discussions about the nature of the threat from Belgian-based submarines and possible tactics to combat it. The RNAS welcomed Navy flyers along on bombing missions. They provided information and analysis to Admiral Sims' London-based Planning Section, made suggestions, and then reviewed Navy proposals to create a northern bombing group, lobbying for acceptance of the Paris-generated plan as opposed to the one advanced by the General Board. Organization and material requirements of the proposed squadrons would be based on those of Handley Page and day bombing units currently in the field.

Once the United States committed to proceeding with the NBG initiative, Britain did everything possible to make it happen. When Cone raised the possibility of American ground personnel training with the RAF in the Dunkirk area, Lambe expressed every desire to help. Cone responded by using his influence in London (via Sims) to secure authority for Lambe to assist the United States.[32] The Northern Bombing Group utilized the legal structure of the British Claims, Hiring, and Requisition Commission to obtain flying fields in the Dunkirk–Pas de Calais region. In July, Cone reported that Lambe had granted the United States "wide privilege for the selection of [NBG] sites."[33]

When the Navy decided to move its proposed assembly and repair facility in northern France away from the impending German threat, the RAF provided an excellent alternate site on the south coast of England. Beginning in June and continuing until virtually the end of the war, RAF instructors, schools, training fields, and operational units tutored hundreds of American aviation personnel in the theoretical and practical methods of war.[34] By midsummer, the Navy had forged an agreement with the Air Board in London whereby American bombing squadrons "will be placed on the same footing in regard to stores and supplies as the Royal Air Force squadrons located in northern France, so that all material used by the two squadrons can be placed in the same storehouse."[35]

Naval aviation headquarters in Paris greatly appreciated this cooperation. Cone complimented the British Supply Division. "This section helps us almost daily in the matter of contracts," he noted. Three RAF officers joined the Navy's aviation organization, the ubiquitous Spenser Grey, Major F. R. E. Davies, and Group Captain Allerenshaw, and "are rendering us invaluable service." Of Grey, Cone reported, "Commander Grey is assigned especially to advise concerning all matters in connection with the [bombing] project." Finally, "cordial" relations had been established with British Air Authorities in the Dunkirk region and Cone opined, "As far as I know our relations have been everything that can be desired."[36] Following the war, Cone recommended Charles Lambe, Spenser Grey, and John Porte be awarded the Distinguished Service Medal for their enormous contributions to the Navy's aviation efforts during the conflict.[37]

For these gestures, British aviation received a great deal in return. At the height of the crisis in April, RAF units at Dunkirk obtained a much-needed infusion of manpower. In June they requested the Navy supply two thousand trained or partially trained ground men, along with necessary officers, to fill vacancies among depleted ranks at their naval air stations. Cone believed the Navy Department should respond favorably, instructing additional personnel in the United States in excess of current needs "for the assistance of our allies." This ran counter to Washington's stated policy of training only enough men to meet Navy requirements. Soon thereafter the British requested the United States supply one hundred pilots, equally divided between those proficient in land planes and flying boats. Two weeks later, Cone urged policy makers to lend all possible assistance toward reinforcing British aviation efforts, indicating tentative approval of the RAF request for pilots, and announcing acceptance of a British offer to turn over the facility at Eastleigh to American use.[38]

Chapter Eleven
Training of Personnel

Training aircrews for proposed bombing squadrons, especially those operating under the cover of darkness, constituted a large, complex undertaking. Lacking facilities of its own, the Navy necessarily relied on "the kindness of strangers," allies and partners in Europe and the United States. This pattern emerged initially in 1917 when the First Aeronautic Detachment and several members of the First Yale Unit received instruction at a variety of French and British schools, including Tours, St. Raphael, Lac Hourtin, Cazaux, Turnberry, and Ayr. Many enlisted observers followed a similar path, training at St. Raphael, Cranwell, Eastchurch, and Leysdown. Balloon and dirigible personnel took classes at Cranwell and Roehampton. Back in North America, a contingent of aviation cadets spent the summer and fall of 1917 in Canada under the tutelage of the Royal Flying Corps.

According to a headquarters tally, acquiring flyers for the Northern Bombing Group required an initial diversion of fifty-six seaplane pilots to the new, land-based initiative. Similar numbers of enlisted personnel attended bombing schools or received instruction at British ordnance and gunnery centers. Though negatively impacting operations at various coastal stations, as well as ongoing efforts in Italy, reallocation of personnel resources reflected Department policy that determined the bombing campaign against submarine bases in Belgium merited the highest priority.[1]

Recognition of the need to instruct personnel in Europe surfaced in discussions by the Paris headquarters Executive Committee held in April, and negotiations with both the British and the U.S. Army soon commenced. Captain Craven, head of the Operations Division, requested that any personnel accepted for instruction be trained in "flights," groups of ten night bombing pilots and assistant pilots and twelve pairs of day bombing pilots and observers, and that the groups be kept together during the training regimen, "working as teams."[2] Captain Cone clearly recognized that instruction facilities might be in short supply, but expressed great optimism based on the quality of men he observed, claiming, "Our general material in the personnel line is so excellent that my experience has been that once we get started to training our people, they are going to make a wonderful showing in a short time."[3]

Serious negotiations commenced in early May, and despite crowded facilities and marked shortages of experienced instructors, the RAF offered to train ten pilots in "aerial bombing, navigation, and Handley Page piloting" at their complex of bases on the Salisbury Plain. A larger number would not be possible, however. As aviation aide Edwards in London explained to Major F. R. E. Davies of the RAF, any additional slots for naval aviators would come at the expense of British pilots, with training schedules filled for the next seven months. "As you will see," he lamented, "I am a bit discouraged along these lines and don't see how we can conscientiously ask to have any more of our people accepted for this work." Nonetheless, he would continue "to feel around in the hope of meeting more success . . . towards having some of our pilots accepted here for training."[4]

In Paris, Ens. George Fearing met with Air Service officials and over a period of several weeks hammered out an agreement to send eighteen two-man crews to the Army bombing school at Clermont-Ferrand. Simultaneous discussions with the French yielded a plan to instruct twenty pilots at Avord and, if necessary, a second group of eighteen pilot-observer crews at Crotoy. A party of twenty pilots at the Navy's Moutchic seaplane school would receive special bombing training so that they might serve as assistant pilots or observers on the first batch of Caproni bombers. The possibility of training personnel in the Dunkirk area constituted an item on Hutch Cone's agenda when he visited the northeast coast in early May. Responding to a request from the Navy Department in Washington for qualified instructors in gunnery and bombing, Spenser Grey prepared a list of three RAF officers for consideration.[5]

The single largest assemblage of naval personnel trained for duty with the Northern Bombing Group consisted of eighteen pairs of pilots-observer/gunners, expected originally to fill the ranks of the first day bombing squadron.[6] Volunteers selected in the spring of 1918 to man new units represented a varied group. Officer pilots included veterans of French service like George Moseley, Fred Beach, Charles Bassett, and David S. Judd. Others received their initial instruction as part of the First Yale Unit, including David Ingalls, Archibald McIlwaine, and Ken MacLeish. A few like Edwin Pou, Woldemar Crosscup, and Arthur Boorse trained first at Squantum, Hampton Roads, Virginia, or at Pensacola, Florida.[7] Many enlisted men selected to become observers/gunners served initially with the First Aeronautic Detachment that reached France back in June 1917, including Randall Browne, Irving Sheely, and George Lowry, all chief machinist mates. Virtually all were combat veterans; many had labored at NAS Dunkirk or flown over the Western Front.

Word of the planned bombing initiative circulated quickly through the service grapevine. In a May letter to his parents, George Moseley discussed the dilemma the project posed. He could choose between staying at Dunkirk with men he admired like Godfrey Chevalier and Di Gates ("I hate to lose my friends"), or opt for "a trip south for a few weeks instruction" and then return to the region as part of a land squadron. Moseley did not seem deterred by any danger he might face. Instead, being stationed at an inland aerodrome would mean living in "very small buildings outlined on the outside with sand-bags," far away from his current "fine quarters and the beach."[8]

On May 10, two officers, presumably from Paris headquarters, came to NAS Dunkirk for dinner, bearing official news of the bombing project.[9] Chevalier then gathered his pilots and asked for volunteers. Virtually all stepped forward. One, Kenneth MacLeish, explained his decision by saying, "I volunteered because we'll never see any action on these seaplanes." High-spirited David Ingalls added, "Hurrah for land machines." Many of the station's enlisted observers signed on as well. Irving Sheely, who served in April with 202 Squadron, RAF, claimed he made the switch because, "I don't fancy a watery grave like four of our number have already gone to."[10] Two weeks later, a general migration from Dunkirk southward commenced. Most men received a short leave in Paris and then continued onward to the U.S. Army's recently opened bombing school, the Seventh Aviation Instruction Center at Clermont-Ferrand.

The U.S. Air Service acquired the facility at Clermont-Ferrand in south-central France in November 1917. According to George Moseley, the school was in a "beautiful location, surrounded by low green mountains, much like the mountains of Vermont. Everything is so quiet and peaceful—quite a relief from the continued bombardment to which we were more or less subjected at Dunkirk." Moseley added that the area had witnessed a fierce battle between Julius Caesar and the Gauls in 45 BC. Originally used by the Michelin Manufacturing Company to test Breguet aircraft, the American Army expanded the facility considerably by adding hangars, barracks, mess halls, a YMCA hut, and other structures. Many were "constructed of white plaster with dainty pink roofs."[11]

Despite a certain amount of friction with Army cadets and instructors, including a spat over the quality of accommodations, Navy flyers soon settled into the camp routine. Up at 5:15 a.m., they flew or attended lectures until 10:00 in the morning, with instruction resuming at 4:00 p.m. after unstable air caused by the heat of the day settled. Work continued until approximately eight in the evening. Enlisted personnel frequently used their off-duty hours to roam the surrounding farmland, picking strawberries from the fields and ripe cherries from abundant orchards. Officers repaired to the resort town of Royat. Headquarters kept an eye on the trainees by dispatching Lt. (jg) Bernard Donnelly from the Ordnance Section on June 11. He remained at Clermont-Ferrand until June 16.[12]

Most everyone commented on the ungainliness of the Breguet 14 B.2 bombers they flew. Kenneth MacLeish called the machines "very heavy and very lazy in the air," while George Moseley compared them to a three-ton truck. Nonetheless, pilot/observer teams acquitted themselves well, earning high marks in bombing, gunnery, and formation flying. With most of their number college athletes, one Navy flyer boasted, "We beat the Army in everything." An outbreak of influenza and a fracas with a dozen Air Service cadets on the last night in camp marred their four-week stay. Following completion of the course, the Navy men returned to Paris and then the Dunkirk region to await temporary assignment to an RAF bombing squadron.[13]

While one group of future bomber crews labored at Clermont-Ferrand, a smaller unit underwent training at the Centre Militaire d'Aviation at Avord (Cher) in central France. These included ten ensigns who took the preliminary course in land flying, lasting about three weeks. They received instruction in the Blériot "Penguin," a simple aircraft with clipped wings making it incapable of flight, quickly moving on to double- and single-control Sopwiths. Following completion of the course, the men transferred to RAF stations in England to receive "practical experience in day bombing."[14] Another plan suggested by Spenser Grey, to train night bombing pilots at Pauillac utilizing Voisin aircraft and then Capronis when delivered, was never implemented.[15]

The first Navy men assigned to Northern Bombing Group night squadrons underwent instruction at a complex of RAF aerodromes established on the Salisbury Plain in Wiltshire, England: Old Sarum (Ford Farm), Stone-henge, and Boscombe Down (Red House Farm). In late April, several offi-cers were detached from the Navy flight school at Moutchic and ordered to England for training, including William Buckhauser, Jesse Easterwood, Leslie Tabor, Hugh Terres, and James Nisbet, all ensigns. By May 9, Cone could report ten pilots undergoing instruction in Handley Pages, with the program to last six-to-eight weeks.[16]

The generally flat, unobstructed terrain on the Salisbury Plain proved ideal for aviation training purposes. Old Sarum aerodrome, about five miles north of the cathedral city of Salisbury and less than 2,000 feet from the Iron Age/medieval ruins for which it was named, opened in August 1917 as home field for three day bombing squadrons created to serve in France. On April 1, 1918, it became home to 11 Training Squadron. A few structures erected in 1917–18 survive to this day. The nearby 320-acre aerodrome at Stonehenge, known officially as No. 1 School of Aerial Navigation and Bomb Dropping, also opened in 1917 to train both RNAS and RAF day and night bomb-ing units. Constructed just southwest of the famed stone circle, it closed in January 1921. The sight of "modern" aircraft like the immense Handley Page 0/400 winging over the ancient megaliths must have startled many. A third facility, Boscombe Down, occupied a location a few miles southeast of Amesbury. It, too, opened in 1917 and closed after the end of the war, only to be redeveloped as a permanent RAF facility in the late 1920s/early 1930s.

By mid-July, ten Navy aviators were undergoing instruction in night bombing at Stonehenge. Six already held orders to report to British squad-rons in the Dunkirk vicinity to obtain practical experience conducting raids and other duties. Moseley Taylor, a pioneer member of the Harvard group of early aviation volunteers, proved such an apt student he was retained for some time as an instructor before being posted to 214 Squadron. The same held true for Bostonian William Gaston, a graduate of the RFC gunnery school at Fort Worth.

Nonetheless, progress for Americans assigned to these schools proved slow, due in part to a seemingly endless series of crashes and shortage of training machines. Long stretches of rainy weather further delayed the pro-cess. Food at Old Sarum was judged excellent (prepared by a U.S. Army con-tingent), but the barracks leaked. The Americans deemed the sustenance provided at Boscombe Down dreadful, and considered obsolete DH-6 training machines employed there "poor and dangerous." A few Navy pilots took a five-week course in DH-9 day bombers at RAF Waddington in Lincolnshire.[17] Back in the United States, forty crews later assigned to the Northern Bombing Group spent much of the late winter and spring at a bombing/gunnery school operated by the Royal Flying Corps at Camp Talieferro near Fort Worth, Texas.

Mastering the huge Handley Page, a true leviathan of the air, required considerable skill. Captain Cunningham, USMC, compared it to a battleship. One British pilot, Laurence Hartnett, called it "immense . . . the biggest thing you could ever think of. You could comfortably walk under the wings." The nose of the aircraft "looked upward to heaven." The engine compartment, the only crew space slightly protected from the weather, contained a few meager instruments, including an airspeed indicator and revolution counter. Rolls Royce "Eagle 8" engines supplied power, "Wonderful engines. Delightful. Well-mannered."[18]

Flying the lumbering beast required a heavy hand. With a 100-foot wingspan and loaded weight of approximately 12,000–13,000 pounds, it lacked aerial agility. Hartnett recalled, "You had a wheel for steering and you leaned into it pretty hard; you were almost on direct cable to work rudder, elevator, and ailerons." The machine tended "to wallow a lot. In other words, they were not sensitive on their controls. To make a turn you pushed on your rudder pretty hard and paused almost before putting on bank, and then you fed the bank on and made a graceful turn very different compared to the scouts and small aircraft." A curved tube with a bubble served as the bank indicator. The large, illuminated compass located between the pilots could be thrown off by swinging the Lewis gun in the nose of the airplane.

Operating at night created many challenges. In the dark, signaling flares— red, green, white—could only be distinguished by "the amount of knurling on the rim." Landing on a clear, moonlit night occurred with the aid of flaming buckets placed on the field and aligned with the wind. Without moonlight, the crew employed Holt landing flares affixed beneath the wings activated by two buttons. They provided about four minutes intense, "magnesium-like" illumination. Once they burned out, however, "you went into dead darkness all of a sudden. One minute you had a fair amount of spread light, next into pitch darkness. . . . Some pilots didn't like using them."[19]

Even with the frustration voiced by many involved in the northern bombing program at the slow pace of development, the Navy and Marine Corps actually made extraordinary headway in a very short time. The decision to implement a bombing campaign was not communicated to Europe until early March and the final form of the effort was not determined until April

30. During the same period, German forces unleashed a titanic battle on the Western Front, threatening to overrun planned staging areas for the bombing squadrons and severely disrupting transatlantic shipping schedules.

Impediments notwithstanding, the four-month period ending in late July witnessed rapid, even remarkable, progress on many fronts. A Northern Bombing Group command structure and staff emerged. Productive liaison was established with the Navy Department in Washington, the U.S. Air Service, the RAF, and relevant French forces. Sufficient personnel were gathered to man four night and four day bombing squadrons. Training commenced, and in some places was completed at facilities operated by the RAF, the French Armée de l'Air, the U.S. Air Service, and Italy's Corpo Aeronautico Militare. Aerodromes were selected and secured and construction rushed ahead. An assembly and repair facility in southern England was secured. Three complete day bombing squadrons crossed the Atlantic, with a fourth to follow shortly. Throughout the buildup, the RAF provided enormous assistance and guidance. Only one component seemed lacking—aircraft to conduct the prescribed missions.

Chapter Twelve
Capronis Coveted

Army versus Navy in Italy

All the plans and progress implemented thus far, whether in France or the United States, would come to naught without the aircraft necessary to conduct offensive operations. Upon this one issue hinged the success or failure of the entire venture. Once the Navy Department reached the decision to proceed with plans to establish a bombing group in Europe, it became abundantly clear to planners in Washington that no night bombers would be forthcoming from America. As early as April 4, 1918, Admiral Sims was advised by cablegram to secure them abroad, if possible.[1] He responded on the 21st, passing on Captain Cone's request for authority to place a contract with the Italians to purchase needed Capronis in exchange for Liberty motors and raw materials.[2]

Cone's involvement with Caproni began late in 1917 when representatives from the firm offered to supply its aircraft to the U.S. Navy in Europe.[3] Plans to obtain bombers for work in Flanders were initiated in February 1918, when Lt. Edward McDonnell, then a member of Cone's staff in Paris, was dispatched to Italy to investigate aviation conditions there, paying particular attention to bombing operations and aircraft.[4] While in Italy, McDonnell met with Signore Caproni on several occasions, toured his factory, and made flights in the industrialist's planes, including a three-engine, 450-hp bomber. Designated the Ca. 3 by the Italian Army, it was the most numerous Caproni bomber in use. During one of his test flights the huge machine—with an Italian pilot at the controls—performed a complete loop. McDonnell rightly considered this to be a remarkable feat for such a large plane. The 450-hp bomber was powered by three very reliable Isotta-Franschini motors. Rated at 150 horsepower, the motor had an excellent record and performed well at the front. The Ca. 3 was highly maneuverable (as indicated by its ability to complete a loop) and capable of carrying a 1,000-lb bomb load to an altitude of 15,000 feet.

McDonnell was so enthused by the plane that he requested permission to fly one from Milan to the front where he spent time with a bombing squadron and even took part in a raid behind Austrian lines, serving as the front cockpit gunner. After returning to Milan, McDonnell met again with Signor Caproni to discuss a new airplane then under development, a 600-hp model powered by three 200-hp Fiat engines that was reputed to have nearly double the bomb-carrying capacity of the Ca. 3 and a top speed of 105 mph.[5]

After returning to France in mid-March, McDonnell was sent to the northern front and attached to 7 Squadron, RNAS, where he flew as a front and rear gunner in combat raids behind the front lines. As a result of this experience, he came away believing that the 600-hp Caproni under development seemed to have advantages over the Handley Page then being flown by 7 Squadron.[6] This information must have been passed to Admiral Sims when McDonnell traveled through London on his way to Washington, D.C., at the beginning of April. And it was certainly forwarded to naval authorities in Washington after he arrived, ensuring the ready acceptance of Cone's proposal to obtain Caproni aircraft for the Northern Bombing Campaign.

With Sims' approval in hand, Cone dispatched Lt. (jg) Harry F. Guggenheim and Spenser Grey—now bearing the rank of a lieutenant colonel in the Royal Air Force—to Rome to negotiate the purchase of Caproni's aircraft with the Italian government.[7] Guggenheim, an experienced businessman, was the son of one of the wealthiest men in America. In 1917, anticipating his country's entry into the First World War, he purchased a Curtiss flying boat and learned to fly. After America entered the war, he helped form a naval aviation unit in Manhasset, New York, and was commissioned as a lieutenant, junior grade, in the U.S. Navy Reserve Force. He served as Cone's business aide until March 1, 1918, when he was appointed director of the Administrative Division in charge of all business and industrial activities.[8] Guggenheim was the logical choice to negotiate the contract and was ordered to Rome to work with Grey, who was already in Italy as part of a commission studying the question of establishing air bases for the U.S. Navy.[9]

Even before their arrival in Italy, the Navy's plan to acquire land bombers for use against the enemy ignited a firestorm at the U.S. Air Service's European headquarters then under the command of Brig. Gen. Benjamin D. Foulois. Relations between the two forces were already strained over a dispute between Cone and Foulois concerning the structure and membership of the Joint Army-Navy Aircraft Committee that had been created to organize and coordinate aircraft procurement activities of U.S. armed forces abroad. What should have been a simple administrative matter became a major issue when Foulois insisted on a majority representation for the Army, despite General Pershing's instructions that three Army and three Navy members be appointed. After the first few meetings in December 1917, it became clear that Foulois was uncooperative, forcing Cone to request dissolution of the committee.[10]

Foulois complained bitterly in April when he learned that 734 scarce Liberty engines had been allocated to the Navy, asking his superiors in Washington why these engines, which were desperately needed by all of the Allied air services, had been given to the Navy.[11] He hit the ceiling several weeks later when he discovered that the Navy was planning a separate bombing offensive against enemy submarine facilities using land planes operated from bases on the Western Front.[12] "Present military emergencies," he wrote in a cable sent to Washington under Pershing's signature on May 3, "demand that the air services of the allied armies be given all priority in advance of the air service of the allied navies. The air supremacy of the Allies on the western front is only held by a narrow margin at the present."[13]

When this cable was sent the Germans' spring offensive was in full swing and presented a grave danger to the Western Front. Although the enemy's offensive toward the Channel ports had been contained, in the center of the Allied line, the Germans still held the initiative. Nevertheless, the War Department, acutely aware of the need to keep the sea-lanes to Europe open, reaffirmed the Navy's bombing measures against the enemy's submarine bases, cabling back:

> Priority to United States Navy Air Services for aviation materials necessary to equip and arm seaplane bases was approved by War Department, November 1, 1917. On March 17, 1918, War Department approved request of Navy Department that 80 two-seater pursuit planes be delivered to Navy on or about May 15, to be used in bombing operations. On May 2, War Department acceded to

request of Navy that this number be raised to 155, but deliveries distributed over longer period. On April 10, War Department concurred with Navy Department that operations against submarines in their bases was purely naval work. Seven hundred and thirty-four Liberty engines have been allocated to Navy for delivery prior to July 1. No allocations have been made after that date. Navy Department for last year has left matter of engine production entirely in the hands of the War Department, and is in this respect wholly dependent for the operations of their Aviation Service. War Department, May 7, carefully reviewed entire matter, in view of your cables, and had decided that no changes can be made at present in priority decision.[14]

Lieutenant Guggenheim arrived in Rome amidst the imbroglio surrounding the issue of the Navy's land planes. One of his first tasks required meeting with Capt. Fiorello La Guardia, the Army's Air Service representative in Italy.[15] La Guardia, a first-term congressman from New York City, had joined the Air Service in the summer of 1917 and learned to fly in Italy at Foggia.[16] This was something of a homecoming for the Italian-American airman, for Foggia was his father's birthplace. In February 1918, La Guardia was designated to represent the Joint Army-Navy Aircraft Committee in Paris with the Italian authorities. This brought him into contact with Lt. Cdr. John L. "Lanny" Callan, Cone's hand-picked representative in Italy. Callan—a native of Cohoes, New York—was an ideal selection for the post having a long and intimate association with Italian aeronautics.[17] While employed by Curtiss, he met Capt. Ludovici De Filippi, who later played a critical role in the Italian Navy's Mediterranean aviation operations. Following the outbreak of war in Europe, Callan sailed to Italy as a Curtiss sales representative. In January 1915, authorities there requested that Callan oversee establishment of their first aeronautics school at Taranto. Curtiss granted permission and in February "Lanny" became chief instructor and assistant to the commandant of the facility. After extensive military service in Italy, he returned to the United States and joined the Navy.

In early March 1918, while plans were being formulated for a bombing campaign against the U-boat bases, Cone asked Callan to make inquiries regarding the Italians' ability to produce Capronis. Callan replied that eight factories were available.[18] These inquiries brought Callan into contact with La Guardia, who was also seeking bombers on behalf of the Army's aerial effort in Italy. Previous efforts to obtain Capronis for the Air Service had been initiated in August 1917 when the Bolling Mission first visited Italy but were never fulfilled in part because the country lacked the materials needed to construct them.[19]

At first Callan assumed he and La Guardia could work together amicably, as both had a vested interest in expediting pilot training and production of Caproni aircraft. In a letter to Cone dated March 27, 1919, he reported being able to utilize La Guardia as an alternate conduit of information from the Commissary General of Italian Aeronautics Eugenio Chiesa.[20] Callan met La Guardia a second time in late March, when the New Yorker urged completion of a naval air station on the Adriatic as a spur to Caproni production. At that time, Callan observed that "La Guardia may be a politician, but he is certainly a very influential one here."[21]

At this juncture, the Navy offered to supply lumber and other materials for its own orders, but La Guardia accused the sailors of playing foul because they controlled transatlantic shipping. He immediately launched a lobbying program that proved highly offensive to Cdr. Charles R. Train, the U.S. naval attaché in Rome. Train informed Paris that his efforts "were frustrated and position greatly embarrassed by the actions of Captain La Guardia," who undercut the attaché by personally interfering with the Commissary General of Italian Aeronautics. The congressman, according to Train's letter, claimed that Navy representatives had no authority to negotiate, nor could they guarantee shipments.[22]

La Guardia played a double game, however. On one hand he did everything he could to sabotage the Navy's efforts to procure Capronis; on the other he tried to placate the Italian authorities by encouraging the Navy's intentions to operate in the Adriatic. Because the U.S. Army had no plans to employ Capronis south of the Alps, the Italians had little incentive to fill the Army's orders for the scarce bombers. If the Navy conducted real combat operations on the Venetian front or elsewhere, however, local authorities might be more amenable to allocating aircraft to the Americans.

Despite the personal animosities that arose between La Guardia and his Navy counterparts, Lieutenant Guggenheim was able to negotiate an agreement—known as the May 10 agreement—with Signor Chiesa that provided for the purchase of thirty aircraft to be delivered in June and July, and eighty more in August, twenty in September, and twenty per month thereafter.[23] In return, the U.S. Naval Forces, Foreign Service would recommend to Admiral Sims that high priority be given to both purchases and shipment of raw material ordered by the Italian government from the United States for the manufacture of aircraft and sufficient raw material would be shipped to Italy by the Navy to compensate for supplies diverted to the planes manufactured for U.S. naval aviation in Italy.[24] The parties never formally signed the agreement, however, and despite several arbitrary modifications made by Signor Chiesa, Italy proved unable to fulfill the terms of the deal.

As the weeks passed there was little movement on either the production or training front. Competition between the services intensified and acerbic charges and countercharges flew back and forth. Cone in Paris, sensing that trouble was brewing, sent word to Capt. Noble Irwin in Washington at the end of May that La Guardia—as a New York politician—was principally concerned in pushing a large general project looking toward a campaign in Congress. He remained decidedly unenthusiastic about working with the Italian-American airman, warning that his people in Rome believed La Guardia had used "underhanded methods" and tried to oppose the Navy's plans to obtain Capronis.[25]

Word of the discord quickly got back to AEF headquarters—perhaps carried by La Guardia himself, who visited Paris in early May and, as the Navy believed, spread false reports of the negotiations with the Italians. In response, Pershing penned a personal letter to Sims describing a "lack of coordination between officers who are handling the U.S. Army and Navy Air Service in Europe." He charged that there were instances when the services had come into competition in obtaining material, creating an unfavorable impression with the Allies that might affect other areas of cooperation. Somewhat taken aback by these charges, Sims wished to respond to the Army chief and solicited Cone's "immediate" opinion.[26]

Cone replied without delay, dispatching a cable to Sims the next day. He had already shown the admiral's communication to Col. Halsey Dunwoody, assistant chief of the Air Service for supply, who asserted, "perfect coordination between our offices exists" in all technical and supply questions. Cone recalled that when he first arrived in France cooperation at all levels had been very good, but recently "it has been impossible to get in close touch with the Chief of the Air Service [Foulois]," another Navy nemesis. He denied that the services competed for material, stating that everything was obtained through the Franco-American Mission and the Army General Purchasing Board. With regard to the Caproni contretemps, the likely source of Pershing's comments was a rat named La Guardia. Cone worried that close cooperation meant letting the War

Department handle the Navy's business, something he did not believe them capable of doing.[27]

Meanwhile, the Air Service experienced its own internal discord, with General Foulois in particular. To remedy this situation, Pershing appointed Brig. Gen. Mason M. Patrick as new chief of the AEF Air Service in late May, moving Foulois to another assignment. Patrick and Cone then sat down to thrash out various issues. Cone offered to let the Army conduct negotiations for Capronis, but objected to La Guardia's involvement. Patrick stood by his troublesome subordinate, but promised the New Yorker "would faithfully care for the Navy's interests." Attempting to put the squabbling behind them, the senior officers drafted a joint memorandum on May 31 addressed to the, Italian Aeronautical Mission Paris, declaring that the Army would represent Navy interests in Rome and they would use their combined efforts to secure raw materials for Italy. Any previous agreements between the Navy and the Italian government remained in force.[28]

Part of the plan to utilize Italian bombers required sending Navy mechanics and engineers to Caproni, Fiat, and Isotta Fraschini factories. Lt. (jg) Austin Potter, the officer in charge of the Fiat and Caproni divisions, spent weeks studying at Italian factories and flying fields, where he subsisted on a diet of pasta, goat milk ice cream, and mutton of "uncertain age but unmistakable flavor." A keen observer of the local scene, he described Fiat shops filled with soldiers returned from the front, those with influential friends, a few youths, and bloomer-clad, black-frocked young women. Work hours stretched from 7:00 a.m. to 6:30 p.m., with one and a half hours off for lunch. Everyone labored on a piecework basis, with maximum earnings of 10 lire per day—equivalent to about $1.35 at that time.[29]

To Potter, manufacturing methods used in the Italian factories seemed primitive. Much of the work was being done by manual labor that would have been performed by precise machine tools in the United States. He decried the inadequate inspection of machining and assembly operations, and was critical of the working drawings prepared by the Italian engineers, which lacked exact tolerances. Instead, the drawings contained symbols, squares, circles, indicating "force fit," "tight fit," "easy/sweet fit," and "running fit," and so forth and so on.

While American mechanics studied in various factories, fifteen Navy pilots led by Lt. (jg) Sam Walker reported to the aviation school at Malpensa in early June. The largest training facility in Italy with perhaps three hundred officers under instruction, Malpensa was situated forty-five minutes north of Milan by electric trolley. The airfield, about three miles square, lay in the stunningly beautiful lake country. During practice flights, students often passed over Lake Maggiore and Lake Como, ascending into the foothills of the Alps. American trainees lived with Italian officers and followed the same rules and regulations as their hosts. Accommodations were clean and comfortable, but the flying schedule differed greatly from that employed in the United States. Instruction began at 5:30 a.m. and continued until 10:30 a.m. Next, everyone sat down to a large meal, followed by a lengthy nap lasting until late afternoon, when instruction resumed, sometimes until ten in the evening. All this was designed to avoid the intense midday heat.[30]

The continued lack of Caproni deliveries, however, cast a pall over the training program. What was the use of learning to fly the things, after all, if none could be supplied. The delivery issue was further clouded by the Italians' reluctance to provide accurate or timely information on when bombers would be available. Unresolved problems limited output so that very few planes were produced, perhaps thirty to forty in June, and forty to fifty in July; far too few to meet the competing needs of all of the Allied air services.[31]

While sensible in terms of presenting a united front with Italian authorities, the May 31 Army-Navy agreement only inflamed suspicions of the airmen of both services stationed in Italy and cooperation proved short-lived. La Guardia, according to Commander Train, rarely visited Rome and proved nearly impossible to contact. Word filtered down that the congressman connived at Malpensa for Army aviators to receive priority training. In a letter to Robert Lovett at Paris headquarters, Callan charged that "La Guardia was profuse in his denials, but Train doesn't believe him." Sam Walker had a particular bone to pick, reporting that "La Guardia was all for the Army . . . letting the Navy twiddle its thumbs." Callan requested that Walker keep him informed, as "it was very important for us to know whether he is working for us or against us." Callan approached the Italian Ministry of Marine and received assurances that the major had been instructed not to interfere. Well aware of the criticism, the embattled congressman attempted to set the record straight with a lengthy communication to General Patrick. La Guardia claimed the Navy acted "so queer and unfair down here," that Captain Cone must surely be misinformed of the conditions in Italy. In a tightly spaced, six-page memo he detailed point and counterpoint of his actions and the Navy's lack of cooperation, duplicity, and misrepresentation. "I need not point out," he concluded, "that the Navy's attitude in this matter was anything but fair and just."[32]

Eventually, the arrival of Maj. Robert Glendenning, La Guardia's replacement, went a long way toward soothing Navy suspicions and restoring a healthy measure of inter-service amity. Though inexperienced, Glendenning proved "a much more satisfactory representative than the former." Nonetheless, the war of memos and charges continued into early August, even after La Guardia relinquished his former duties. Callan sent a personal communication to Cone in Paris, including much of his correspondence with the Army representative, "together with memoranda I made on La Guardia's answer." In several places, Callan included underlines and special notes for emphasis, "In order that you may see for yourself exactly how he acted down here and know that he has not represented us in the proper manner or worked for the best interests of the service." Under no circumstances, Callan declared, should the Navy deal with La Guardia. The New York congressman, it seems, had left everybody in the lurch, including his own replacement, and his actions in the final two weeks of his tenure "were absolutely intolerable."[33]

Unfortunately, belated cooperation between the Army and Navy did nothing to speed construction of critically needed Caproni bombers or remediate the severe defects that soon became apparent in the small number handed over to American forces. The decision to rely on Italy as the sole source of crucial aircraft proved to be a fatally flawed one from which the Northern Bombing Group never recovered.

Chapter Thirteen
Airbases and Support Facilities

C reation of a large industrial facility to assemble, test, service, and
repair the Navy's bombers stood near the top of the list of crucial
needs to be met before squadrons could take to the skies. Original
plans envisioned establishing a site in the Pas de Calais region near proposed American aerodromes. Initially designated Base "B," the assembly and
repair complex would be responsible for assembling and testing all aircraft,
repairing aircraft and engines, supervising aerodrome construction, hold-
ing reserves of all stores and equipment, and supplying squadrons with fuel,
bombs, and ammunition. Staffing was estimated at approximately two thousand officers and men.[1] In April, Lieutenant Lovett headed north to locate
a site for the "repair base field." At the same time, Cdr. Benjamin Briscoe
readied for a trip to the United States to organize the men and equipment
necessary to operate that establishment.[2]

Success of the German spring offensives raised fears on both sides of the
Atlantic that the anticipated facility might be overrun, eliciting a decision
to abandon the existing plan and seek a safer location across the Channel.[3]
According to Cone, the ubiquitous Spenser Grey "who is here with me, is
keeping me informed of every move so that we will not only be able to move
to England at a moment's notice, but will also be able to occupy aerodromes
if it becomes necessary."[4] Charles Lambe—now a general officer in the
RAF—agreed to make a repair depot at Guines available to the Americans,
while the British promised to identify several possible sites in England.[5]

Further discussions yielded a June 21 offer by the Air Ministry to turn
over a partially completed acceptance depot at Eastleigh, near Southampton.

139

A party led by Captain Hanrahan traveled to England at the beginning of July to inspect the proffered facility. They observed much work in progress, with hangars approximately 90 percent complete, storehouses 30 percent finished, and living quarters for 300 men already in place. Hanrahan returned to Paris a few days later and reported "at length" on the site, Executive Committee notes indicating "large portion of meeting was devoted to description by Hanrahan of repair base and field Eastleigh."[6]

Authorization to accept the site followed quickly and Sims could report July 16, "Have practically completed negotiations to take over the acceptance park at Eastleigh."[7] Civil Engineer Frederick Bolles hurried down from London with orders to report to Hanrahan. By July 17, the British offer had been officially accepted and the Navy assumed control three days later. Lt. Godfrey Chevalier was given the task of readying the facility for use, assisted by engineer Bolles. Captain Cone exulted in this turn of events, calling acquisition of the Eastleigh site a "bonanza, . . . in every way admirable for our needs."[8] By the end of the month, Paris reported efforts under way to divert activities from Pauillac to Eastleigh, with tools and materials being shipped to England.[9] It was hoped these measures would allow DH-4s transported from the United States for northern bombing use to be assembled just across the Channel rather than in distant southwestern France.[10]

Even as the task of creating an efficient repair and assembly base at Eastleigh proceeded, the work of establishing aerodromes and other facilities in France moved ahead. In early April a group consisting of Craven (Operations), Briscoe, Billings (Civil Engineers), and Spenser Grey traveled to the Pas de Calais region to inspect possible airfield sites. They returned to Paris April 12, having made some progress, though the heavy fog that frequently blanketed the area hampered their efforts. The following week, Lovett, back in Paris after his stint with the RAF, briefed the Executive Committee on three fields he had identified for possible use. Lieutenant Guggenheim reported negotiations under way for one field even as he gathered information preparatory to making a second application. He was directed to move on the issue as quickly as possible. The process continued into early May, with Cone and engineer Billings visiting Dunkirk to speed the work along. At the same time, Hubert Burnham, a noted Chicago architect, moved up from the Navy's principal supply base at Pauillac to visit the region and make recommendations.[11]

Complicated negotiations involving three countries, several local and regional authorities, numerous layers of command, and a host of individual landowners proceeded slowly. Proposed fields would be acquired through the good offices of the British Claims, Hiring, and Requisition Commission. By mid-May, the Navy had presented applications for four airfields, but Vice Adm. Pierre-Alexis Ronarc'h, French regional naval commander, claimed he received instructions to proceed on only one. After four more weeks of negotiations, exchanges of documents and information, pressure, and cajoling, the bulk of the work was done. In June, Lt. Cdr. H. A. Allen returned to Paris from Dunkirk having completed his duties "in connection with the lease of land for the Northern Bombing Squadron."[12]

Based on the projected size and operational tempo of the Navy's original bombing program for twelve squadrons, initial planning identified six sites. The first proposal designated St. Inglevert, about six or seven miles southwest of Calais, as headquarters and repair base. A new airfield at Campagne-lès-Guines, four miles south of Calais, would become home to Night Squadrons 1 and 2, with Squadrons 3 and 4 occupying a field at Sangatte. Squadrons 5 and 6 would make their home at La Fresne. Marine day bombing Squadrons 7, 8, 9, and 10 were slated to utilize a new aerodrome at Oye between Gravelines and Calais, and Squadrons 11 and 12 would be stationed at Spycker, just south of Dunkirk.[13]

Following Department-mandated reduction of the program at the end of May, Paris juggled some of these dispositions. Hanrahan and his staff

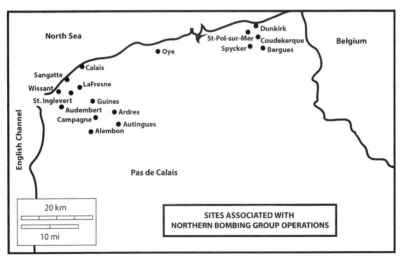

Supporting the Northern Bombing Group program required the establishment of a large network of American and Allied fields and facilities.

would occupy a modest headquarters complex at Autingues, with St. Inglevert now serving as home to Night Squadrons 1 and 2, and Squadrons 3 and 4 ensconced at Campagne. Day bombing Squadrons 7 and 8 continued at Oye, with Squadrons 9 and 10 now at La Fresne. The Navy derequisitioned sites at Spycker, Sangatte, and Alembon. Campagne never conducted flying operations. Damp conditions at Oye resulted eventually in all Marine aviation activities being concentrated at La Fresne.[14]

Having secured the use of future flying fields, the Navy began moving men and material to various locations, initiating the process of turning open agricultural land into aerodromes. On June 23, a pioneer party of Night Bombing Squadron 1, commanded by Ens. Clarence Johnson and consisting of six officers and seventy-four enlisted men in a convoy containing eighteen large trucks hauling construction materials, tools, and other supplies, made their way from Pauillac to Paris to St. Inglevert. On high, open ground they quickly set to work erecting tents, digging small bombproof dugouts, and clearing and leveling the fields. Other work parties soon followed—fifty men from Paris on July 6, sixty more ten days later. The Navy rented a nearby chateau and installed a deep well pump to supply water to the headquarters and field. From the British, they obtained a large, steel frame hangar measuring 80 ft. × 180 ft., which they erected on elevated foundations to permit access for Caproni bombers. Bluejackets also raised a canvas Bessaneau hangar moved from the base at Dunkirk. A squadron of RAF Handley Page bombers occupied a portion of the field, having recently been blasted out of their facilities at Coudekerque.[15] Though harvesting crops temporarily delayed some construction, use of prefabricated wooden structures and strong American backs allowed the process to move quickly once local farmers completed their chores. In time, facilities for six hundred men appeared. Flight personnel also began arriving, including most of the observers fresh from their training at Clermont-Ferrand.[16]

During the summer, work began on the second planned night bombing aerodrome at Campagne-lès-Guines. Here, too, the need to wait until farmers harvested crops delayed progress, though by mid-July construction was under way. For several weeks, the men lived in tents due to delays encountered shipping portable buildings from Pauillac. Several more weeks passed before erection of hangars commenced. Work on these large buildings continued, in fits and starts, until the end of October. Adjacent to the field, the Navy leased a second chateau and built additional quarters for officers with material obtained from the British.[17] The original plan envisioned night bombardment wing headquarters relocating here from St. Inglevert when the first two squadrons were established, but work never progressed that far. Just a mile from Campagne aerodrome stood the RAF Repair Depot at Guines, the use of which had been extended to the Americans. During the summer, material to accommodate four hundred Navy mechanics was placed on order, sufficient to build additional huts, hangars, and store-houses. Construction continued throughout the fall, usually carried out by crews dispatched early each morning from Campagne.[18]

By late July, plans were finalized to locate Hanrahan's NBG headquarters at Autingues (Ardres), about eight miles SSE of Calais and within easy distance of aerodromes at St. Inglevert and Campagne. He and his staff shifted from St. Inglevert in August. Located in a lightly forested area, the site accommodated the headquarters complex, motor transport park, and central supply depot. Construction crews quickly erected prefab structures obtained from the British and the Navy's supply center at Pauillac.[19]

Work at Oye, first of the day bombing aerodromes and intended originally to accommodate four Marine aero squadrons, commenced in early summer, though hampered by the same agricultural delays experienced at the night bombing fields.[20] Located between Calais and Gravelines, only a mile from the coast, the aerodrome at La Fresne later supplanted this field. The RAF supplied Bessaneau hangars. Tents provided shelter until more substantial structures could be obtained from Pauillac. Navy crews and officers initiated work, soon replaced by Marine squadrons, which reached France in late July and arrived at Oye a few days later. By early August, ten officers and four hundred men occupied the raw site.

David Ingalls and David Judd, Navy pilots just returned from combat service with 218 Squadron, labored at Oye for a few days in late July. Ingalls told his father, "I have to reap hay, board ditches, erect hangars, huts, etc. . . . we felt low yesterday when we arrived at a beautiful field of oats, or alfalfa, or some darn vegetable, and put our luggage in one of the few tents set up in a muddy corner." Ingalls seemed consoled by the thought that "the harder we work now the sooner we'll be able to fly and bomb."[21]

At the same time work began on aerodromes and other facilities in the region, the first Marine Corps aviators reached Europe. Lt. Edmund Gillette Chamberlain sailed for France in late May, reporting to Hanrahan in Paris June 5. He had been selected to go abroad and collect material on the other side, being the only officer who could be spared who was familiar with the work. Chamberlain traveled to Dunkirk, "reporting to Captain Lambe for special duty." In July he visited Clermont-Ferrand and Pauillac before joining the Day Wing of the Northern Bombing Group, from which he was detailed to 218 Squadron.[22]

In mid-July, a larger group of leathernecks reached Paris, whereupon they received orders to decamp for Dunkirk where they served with the British for a few weeks, familiarizing themselves with day bombing operations.[23] This advance party for the First Marine Aviation Force to follow included designated squadron commanders Russell Presley, Roy Geiger, Douglas Roben, and William McIlvain. Assistant quartermaster Robert Williams, an intelligence officer, and additional staff and enlisted men rounded out the unit. They were charged with learning the lay of the land, negotiating to arrange training with the British and French, and preparing the way for three squadrons of Marines just then boarding their transport at the Hoboken docks. McIlvain served temporarily as an observer with 217 Squadron.[24]

NAS Dunkerque, the Navy's principal aviation base in France for combating the German U-boat menace, proved ill-suited for the task. Restricted and dangerous operating conditions greatly limited the site's effectiveness, while continuous bombardment from enemy ground, aerial, and naval forces often inflicted severe damage, seen here. The use of flying boat bombers to attack submarines would never defeat the threat. Something new and daring would be necessary. (National Archives and Records Administration [NARA])

Capt. Hutchinson I. "Hutch" Cone (1871–1941), former chief of the
Bureau of Steam Engineering, commanded naval aviation forces in
Europe during World War I and proved to be one of the strongest
proponents of aerial bombing. (NARA)

Wing Commander Spenser Grey (1889–1937) of the Royal Naval
Air Service/Royal Air Force was a pioneer of aerial bombardment,
carrying out his first missions in the opening months of World
War I and playing a pivotal role in developing Britain's bombing
campaign through the next two years. First assigned to assist the
nascent United States Air Service bombing program he soon joined
the U.S. Navy effort instead. (Fleet Air Arm Museum)

Robert A. Lovett (1895–1986) was the son of the president of the Union Pacific Railroad and a charter member of the First Yale Unit. His research and advocacy were instrumental in developing the Northern Bombing Group program, and he later served as commanding officer of its Night Bombing Wing. During the war he rose from the rank of ensign to lieutenant commander. (Naval History and Heritage Command)

Lt. Edward O. McDonnell (1891–1960), shown here in a formal portrait wearing the Medal of Honor he earned for valor at Vera Cruz, played a crucial role advocating and organizing the Northern Bombing Group. His work carried him from the United States to England, France, and Italy. (NARA)

Maj. Alfred A. Cunningham (1881–1939), the "Father of Marine Corps Aviation," shown here in Miami in the spring of 1918, labored tirelessly to promote aviation in the Corps and secure a place for leatherneck flyers in the anti-submarine bombing campaign. His report to the General Board played a major role in convincing that group to support a bombing initiative. (Archives and Special Collections—Marine Corps Library)

A heavy bomber was needed to carry out proposed raids in Europe, so much effort was expended to implement production of the Caproni Ca.3 aircraft in the United States. Sadly, technical and logistical impediments intervened, little was accomplished, and the Navy was forced to seek aircraft elsewhere. (NARA)

Marine aviators underwent their final instruction in the United States at the former Curtiss school field on the edge of the Everglades. The famed Curtiss JN-4 "Jenny" served as their principal training machine. They arrived in crates and were assembled by Marine mechanics at the field. (Archives and Special Collections—Marine Corps Library)

Aviators learned the rudiments of acrobatics, gunnery, and formation flying at the Marine Flying Field near Miami. When the men departed for Europe in July they were full of enthusiasm but lacked extensive or realistic training. (Archives and Special Collections—Marine Corps Library)

Desperate to secure aircraft to equip the Northern Bombing Group, the Navy entered into an agreement with the Italian government to secure scores of Caproni Ca.44/Ca.5. Production difficulties at the factories, however, severely limited production. Aircraft actually delivered incorporated fatal structural flaws and dangerously unreliable Fiat motors. (Naval History and Heritage Command)

Navy pilots destined to crew Caproni bombers in combat received training in Italy. Ens. Alan L. Nichols (right), shown here in the summer of 1918, died on August 17 at Mirafiore during efforts to ferry aircraft from Italy to northern France. (Naval History and Heritage Command)

The German U-boat complex in Belgium was centered in Bruges, an inland port connected to the North Sea via canals to Ostend and Zeebrugge. These concrete sub pens sheltered vessels during rest and refit between cruises. The Allies tried repeatedly to destroy them, but with little success. (Naval History and Heritage Command)

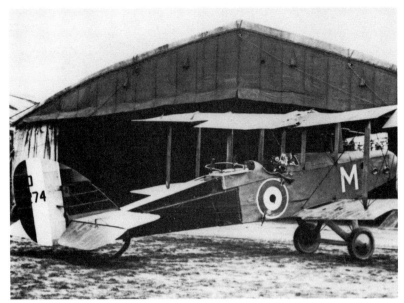

Navy pilots trained to carry out daylight raids were instructed first at the U.S. Army's bombing school at Clermont-Ferrand and then attached to RAF units for a few weeks to receive battlefield experience. Many served with the No. 218 Squadron, flying slow, underpowered DH9 machines. This example, much the worse for wear, made it back to the field despite serious damage inflicted by deadly accurate German anti-aircraft batteries. (Courtesy Roger Sheely)

Navy flyers also served with RAF units operating huge Handley Page 0/100/400 bombers. Following a mission conducted by the No. 214 Squadron on the evening of September 24, Lt. (jg) Alexander McCormick, pictured here during his early flight training, died, struck by the aircraft's starboard propeller after alighting from the rear cockpit. (Courtesy Peter Mersky)

The Navy established a sprawling assembly and repair facility for its Northern Bombing Group aircraft at Eastleigh on the south coast of England. Obtained from the RAF, Eastleigh developed into one of naval aviation's largest logistical centers. (NARA)

Land acquired for Northern Bombing Group aerodromes usually consisted of farm fields in active use. In some cases, the Navy waited for harvesting to be completed before beginning operations. This view of St. Inglevert shows some of the facilities erected to house and feed the hundreds of personnel who manned the base. (Naval History and Heritage Command)

As naval aviation's most important offensive initiative, the Northern Bombing Group attracted a great deal of scrutiny. This photo taken in August 1918 shows commanding officer Robert Lovett in discussion with members of the House of Representatives' Naval Affairs committee who conducted an inspection tour in the late summer of that year. (Naval History and Heritage Command)

Capt. David C. Hanrahan (1876–1944) established Northern Bombing Group headquarters in a wooded site at Autingues in the Pas de Calais region, about ten miles southwest of Calais and the coast. Navy crews used pre-fabricated structures to erect the facilities in just a few weeks. (NARA)

Wing Commander Spenser Grey played a crucial role organizing the Northern Bombing Group, providing both expertise and liaison with the Royal Navy and Royal Air Force. In many ways he proved to be the indispensable man. According to Marine major Alfred Cunningham, Grey was actually running the entire show. He is shown here, seated second from the right, at NBG headquarters sometime in the late summer or early fall of 1918. (NARA)

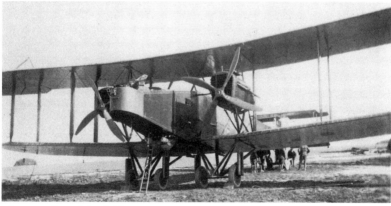

The Navy shared the field at St. Inglevert with the RAF, which operated enormous Handley Page bombers. The British had been bombed out of their aerodrome near Dunkerque and relocated to the comparatively safer environs of the Pas de Calais. (Naval History and Heritage Command)

The Navy facilities at St. Inglevert eventually comprised a significant complex of structures set down in the midst of farm fields. Here horses and a farmer's wagon commandeered for military use share space with an enormous hangar and one of the few Capronis to make it safely to France in August and September 1918. (Naval History and Heritage Command)

Caproni bombers, acquired with difficulty and ferried to St. Inglevert at great cost, proved virtually unusable due to structural weakness and dangerously malfunctioning Fiat motors. (Naval History and Heritage Command)

Marine Corps aviation ground crews manned the day bombing units of the Northern Bombing Group. Here a trio of ground personnel join forces to spin the prop of a DH-4 machine based at the aerodrome at La Fresne. (Archives and Special Collections—Marine Corps Library)

This group of Marine flying personnel, both pilots and observers, pose before their DH-4 aircraft at La Fresne in the fall of 1918. The observer-gunners seated on the ground proudly carry their deadly Lewis machine guns. (Archives and Special Collections—Marine Corps Library)

The day squadrons of the NBG conducted numerous bombing missions in September and October 1918, especially targeting German rail lines and yards, attempting to disrupt enemy efforts to bring reinforcements or supplies forward. (Archives and Special Collections—Marine Corps Library)

Only a few weeks after performing the aerial heroics that would earn him the Medal of Honor, Lt. Ralph Talbot of Weymouth, Massachusetts, died in a fiery crash at La Fresne aerodrome while attempting a test flight on his DH-4 aircraft. The machine failed to lift off completely and the undercarriage hit a berm protecting a bomb dump located near the end of the field. Not wearing his seat belt, passenger Lt. Colgate Darden was ejected from the crashing aircraft and flew far from the resulting inferno. Though he suffered severe injuries, he survived. (Archives and Special Collections—Marine Corps Library)

Capt. David Hanrahan, commanding officer of the Northern Bombing Group, is pictured with his headquarters staff in an informal group photograph taken at Autingues, most likely around the time of the Armistice. Hanrahan had earlier gained fame for captaining the Q-ship *Santee*. (NARA)

NAS Killingholme on the Humber estuary in northeastern England developed into the Navy's largest operational aviation base in Europe. It would have been the center of a joint USN-RAF program to launch heavy bombing seaplanes from lighters towed to within a few miles of the German and Dutch coasts. The program was canceled in August 1918 over questions concerning its likely success. Personnel pictured here are perfecting their machine-gun aiming skills. (NARA)

John Lansing Callan (1886–1958), a pilot for the Curtiss company before the war, amassed considerable experience working with the Italian Navy and for most of 1918 directed the U.S. Navy's aviation efforts in the Mediterranean. He strenuously advocated expanded bombing operations and creation of a Southern Bombing Group that would have exceeded the effort under development in northern France. (Library of Congress)

The sea sled concept, advanced by Cdr. Henry C. Mustin (1874–1923), was designed to transport and launch heavy bombers from the water, thus allowing them to attack German fleet anchorages at Kiel, Wilhelmshaven, and elsewhere. While attempts to launch a Caproni bomber proved unsuccessful, the Navy did manage to get a smaller aircraft airborne in March 1919. Patrick Bellinger called this the Navy's first aircraft carrier. (Courtesy Colin Owers)

Chapter Fourteen

Send in the Marines

E ven as Navy crews trained in France, plans promulgated in Washington assigned Marine Corps aviators responsibility for manning
the day bombing component of the planned assault on the German
U-boat complex in Flanders. Quickly creating battle-capable units would
thus prove essential to the overall success of the entire initiative. But training
and equipping sufficient manpower posed a tremendous challenge. Prior to
the outbreak of war, only a handful of leathernecks secured flight instruction. Two received Army training in land planes. The total cadre of enlisted
personnel numbered barely thirty. Stationed at Pensacola, they formed a
tiny "Marine Section, Navy Flying School."

As war clouds gathered in the winter of 1917, officials in Washington
took tentative steps toward meeting the expected challenges. In late February,
the Marine Corps began organizing an Aviation Company for the Advanced
Base Force (a Marine expeditionary force) at the Philadelphia Navy Yard.
Congress' April declaration of war accelerated events dramatically. With
Capt. Alfred Cunningham leading, the Corps expanded its aviation program
dramatically, recruiting additional personnel, seeking missions, and negotiating with the Army and Navy for equipment and facilities.[1]

Karl Day, a future Marine general, had just arrived at the Officer School at
Quantico, Virginia, in mid-August 1917 when his commanding officer called
the men together and announced that Captain Cunningham from headquarters had a message for them. Cunningham stepped forward and stuttered,
"Uh, uh, Gentlemen, the Secretary of the Navy has a-a-a-a-authorized
t-t-t-t-the" (that is where he got the nickname mutter [mother] Cunningham).

Undeterred, the pioneering aviator explained that the Corps was forming a flying unit to go to France, and that "Anybody who is interested take two steps forward." Day later recalled that most of the officer candidates stepped forward. According to Day, "We weren't a high society bunch, but had a lot of athletes. We were looking for adventure." When Officer School ended a few months later, eighteen men received orders to report to the first Aeronautic Company forming at Philadelphia.[2]

Another future general, Francis Mulcahy, offered a somewhat different perspective. When first offered the chance to enter aviation, he declined, saying he did not want to be a daredevil. He joined the Marines instead and soon went to a rifle range in Maryland where he took some bayonet practice. At a strapping 135 pounds he thought, "What chance have I got if I run into some big Heinie over there. I sat down and wrote an application to aviation. I never heard a word from it." Later, Cunningham came down to Quantico. Of the 300 to 350 officers training there, at least 150 went to see him. "He had a little table set up in one of the barracks, people lined up outside all day long." Mulcahy did not think his short interview went well, but at the end of course orders came through to report to aviation.[3]

Awaiting news, Karl Day told his parents that he expected orders any day and would soon take a very stringent physical examination. He expected to receive theoretical and mechanical training at Philadelphia, followed by work at an Army flying school, likely San Antonio, then maybe at Miami. After qualifying, Day presumed he would be sent to a French facility and taught trick and acrobatic flying. He noted that Captain Cunningham expected to go to France in the next few weeks, to observe and study, and remain for two months.[4]

As Cunningham interviewed potential recruits, two prospective missions for the Marines emerged: anti-submarine patrol duty and reconnaissance/artillery spotting. Piloting seaplanes, the men of the Aviation Company then forming at Philadelphia, reinforced by new recruits and men transferred from Pensacola, would execute the patrol mission. The Corps' decision to send an infantry brigade to France elicited Navy Department permission to form a second flying unit, equipped with land planes, to carry out necessary reconnaissance and artillery spotting assignments. Original plans envisioned a single unit of 11 officers and 178 men equipped with 12 aircraft and 4 kite balloons. The Army would supply both training and equipment. Writing home in early October, Brooklyn native Ed Kelly reported that neophyte Marine aviators in Philadelphia had just been spilt into four companies: observation, combat, coast patrol, and balloon. Scuttlebutt said the first was heading for France, the second to England, and the balloon men to Nebraska.[5]

Even this modest expansion of aviation capabilities necessitated strenuous recruitment efforts, but by mid-October the Aeronautic Company had grown to 34 officers and 330 enlisted men. Utilizing two Curtiss R-6 floatplanes and a single Farman land craft, they received limited flight training. On October 14, the Corps took definitive steps to implement earlier plans establishing the 1st Aeronautic Company of 10 officers and 93 men to carry out the anti-submarine mission and the larger 1st Aviation Squadron of 24 officers and 237 men to handle infantry support assignments in France.[6] The 1st Aeronautic Company soon relocated to NAS Cape May in New Jersey to continue training, and in January 1918 sailed for the Azores where they conducted anti-submarine patrols until the end of the war, returning to the United States in January 1919.

To secure proper training for the 1st Aviation Squadron, Cunningham negotiated an arrangement with Col. Henry "Hap" Arnold of the Army Signal Corps whereby Marine aviators would receive basic flight training at Hazelhurst Field on Long Island before undergoing more advanced instruction at facilities near Houston, Texas. As soon as the Marine squadron was ready for duty in France, the Army would "completely equip it with the same technical equipment furnished for their squadrons." Marines under the command of Capt. William McIlvain began transferring from Philadelphia to Long Island flying fields October 17.

The instruction center on Long Island occupied a 1,000-acre expanse on the Hempstead Plains in Nassau County. A 15-foot-high bluff bisected the site in a north–south direction, creating two distinct sections. Established in 1916 to train National Guard pilots, the Army occupied the entire property in 1917, naming it Hazelhurst Field after Wilson Leighton Hazelhurst, a West Point graduate who died in an aviation accident in 1912. The government acquired additional land south of the original site, forming a large rectangle. The southeast portion eventually became Mitchel Field, and the southwest quadrant, Camp Mills, an Army encampment. Of the original aerodrome, the western (lower) portion was later known as Curtiss Field and the eastern (upper) as Roosevelt Field. It was from the latter area that Charles Lindbergh took off for his 1927 transatlantic flight.[7]

Following their arrival on Long Island, Marine flyers occupied a position along the northern edge of Hazelhurst Field where they lived in 8 ft. × 9 ft. walled tents with no floors or electric lights, but containing a canvas bunk, lantern, and oil stove. Reveille jolted everyone awake at 5 a.m. About 200 enlisted men accompanied the pilots. One enlisted Marine took umbrage at a magazine photograph that identified their camp as the Army's. "Anyone would know," he sniffed, "that the army could never lay out tents as straight as those in the picture." Karl Day believed the program would last about six weeks, combining ground and flight school. "Just watch us," he boasted, "And remember, we are MARINES!" Knowing cold weather was on the way, Day ordered more blankets and four suits of woolen underwear. Captain Cunningham had already said good-bye to the men, having headed off to France "to get the latest wrinkles."[8]

Fledging aviators received instruction in Curtiss JN-4B "Jennys" with deperdussin controls. This system consisted of a U-shaped yoke that extended upward from the cockpit floor on each side of the pilot's legs and then across in front of him. In the center of the horizontal portion of the yoke was affixed a wheel. The entire contrivance hinged backward and forward and controlled the aircraft's elevators, while the wheel moved the ailerons. The pilot worked the rudder bar with his feet.[9] Though the Army allocated space for the Marines, for the most part they ignored the leathernecks, who hired civilian instructors.

Day recalled that some of the instructors were very good, and others quite indifferent. Each took charge of four or five neophyte pilots. One instructor, a certain Kellanan, seemed scared to death and refused to let his students handle the controls. After five flights, Day had still not touched the throttle or rudder. "We raised hell about him and he got fired."

Day's next instructor, James R. Doolittle, went to the other extreme. He was described as "an American who flew over the trenches with the Lafayette Escadrille, but no longer fit for fighting flying." After less than two hours dual flight time, Day watched Doolittle "stagger out of the plane drunk as Hell, nearly put his foot through the wing. He said, 'All right you sons of bitches, go up and kill yourselves.'" That was Day's cue to take off for his first solo. Other students had better experiences with Doolittle. Francis Mulcahy called him a good man and a good instructor.[10] Day nicknamed his fellow trainees the dirty dozen, not too crazy, not too conservative, a sound group of aggressive young men.

The fall of 1917 offered much good weather and the young aviators made rapid progress, with one correspondent noting, "All the new lieutenants are getting along well with their flying and some of them are flying alone. The weather here has been splendid for flying except 2 or 3 days."[11] Most found the Jennys easy to fly, but awfully slow, "you didn't go very far very fast, but they were good training planes." Students took their flight tests at the end of the fall, qualifying as Reserve Military Aviators.

In December, however, temperatures plunged, on one night to 17 below zero, with the men still living in tents. Flying activity ceased December 12. Students would hop the train for New York City "for no other reason than to get warm, get a hot bath and get warm." Many hotels offered servicemen a 25 percent discount. Threats to health became a major concern and the squadron medical officer strongly urged a change of camp. Shortly thereafter, Captain McIlvain ordered his men and equipment loaded on board a train headed south. Though some veterans claimed the group set off without an identified destination, the complexity of securing scarce rail transport for such a venture makes such a scenario unlikely.[12]

While training on Long Island proceeded, the Marine Corps on December 15 established a third aviation unit at Philadelphia designated the Aeronautic Detachment under the command of Capt. Roy Geiger. Initially consisting of only four officers and thirty-six enlisted personnel, mostly drawn from the 1st Aviation Squadron, its mission remained unclear, though it seems to have been formed to cooperate with the Advanced Base Force. It was at this point that plans formulated by the Marines began to unravel. In late fall and early winter, Captain Cunningham made his extended inspection tour of American and Allied aviation facilities in Europe. According to the Marines' official account, Cunningham labored unsuccessfully to convince the Army to attach his flyers to their own infantry brigade as originally envisioned. This rebuff, of course, led to discussions with the Navy in Paris and Dunkirk and, ultimately, his recommendation to the Commandant and General Board that Marines initiate/participate in a major anti-submarine bombing campaign.

The shivering refugees from Hazelhurst Field passed through Washington, D.C., in early January 1918 and proceeded southward to Lake Charles, Louisiana, site of the Army's Gerstner Field. For many it was their first trip through Dixie. Karl Day told his parents, "We had a dandy trip down." Brooklyn's Ed Kelly called it a "regular tour." At longer station stops, he conversed with the locals, whom he found interesting and neighborly. "Nearly all of the Alabama towns and those of Louisiana look poverty stricken. In some it would remind you of the west, one story buildings, a saloon with a false front, a grassless stretch for a road, cows and razorbacks wandering everywhere."[13]

The Marine flying contingent reached Gerstner Field in early January. Since June the previous year, Lake Charles citizens had worked to acquire an Army training camp for their area. In August, they succeeded in convincing the Air Service to establish a 2,000-acre flying field 18 miles southwest of town. Over the next year, the government erected 24 hangars, 24 barracks for officers and enlisted men, a dozen mess halls, 4 warehouses, and attendant offices and shops, all painted green with white trim. Eventually the field accommodated 3,000 men, the first arriving in November 1917.[14]

Taking in their new surroundings, the Marines noted the vast flatness of the place, what one called "the large range of vision." "The country is so flat," he observed, that "you can see the [setting] sun disappear like a nickel in a bank. The evening haze on the ground colors up and looks like a pink gas attack."[15] A young flyer described the land as "flat as a table, only 15 feet above sea level. You can see for miles and miles without seeing a dozen trees." The wind roared across the treeless "prairie," but the weather was beautiful (initially), just like May. The "wrath of nature," soon banished the balmy spell, however, bringing rain, sandstorms, and hail. Temperatures dropped to 40 degrees and wind whistled through the barracks. At first, the base commander refused to house the Marines, but upon receiving proper documentation, provided necessary quarters and integrated the leathernecks into the instructional program. Accommodations proved satisfactory, with hot and cold showers, "eats fine, and lots of sugar."[16]

The Marines went right to it, even laboring on Sundays, what one called "ungodly hours."[17] After assembling their own training machines, they commenced flying, utilizing a variety of aircraft, including Curtiss JN-4C "Canucks," Curtiss JN-4Ds, and Thomas-Morse S4-C trainers. The "Canucks" had smaller tail surfaces than B or D models and came equipped with stick controls. Victor Vernon served as chief instructor, a man Karl Day described as "one of the grandest gentlemen I've ever known in my life."[18]

Work commenced at six in the morning and the men knocked off twelve hours later. One officer recalled, "We ran our own show. The Army had made an awful mess down there. They soon had Marines running every hangar. The Army had no airplanes in flying condition. Our Marine mechanics put their show on the road." Expressing pride of Corps, the young aviators could not help bragging that they filled all the important jobs at the field, even to the exclusion of the Army sergeants. "Of course the dispositions of the army men have not improved under this regime, but the Marines do not fret." In time, a Marine assumed supervision of each aircraft, with "three low forms of life known as army men" under their charge. The results, of course, were "stupendous," with the Army types "beginning to absorb some of the Marine spirit." One day, 47 aircraft took to the skies, amassing 247 hours of flight time. "The best the army ever did was a hundred and three hours."[19]

Keeping the planes airworthy required considerable effort. Ed Kelly compared the machines he worked on to his father's Ford that had to "be coaxed to talk." Blowing sand interrupted flying and damaged engines. Chronic parts shortages idled many men and machines. According to Kelly, the motors were in bad shape because "they have as yet no way of repairing them. Extra parts are expected sometime, but at present we do the best we can." Liberty motors, however, elicited praise, with one mechanic gushing, "All I can say is that they are wonderful." When new aircraft arrived, workers hauled the old ones out to a corner of the field and abandoned them.[20]

Actual flight operations encountered further obstacles. The field was lumpy and boggy, with good landing spots hard to find. On a single day in early February, a dozen aircraft foundered while alighting when their wheels sank into the soft ground, causing them to flip over. Luckily, no one was hurt. During cross-country flights, many planes failed to return, sometimes due to accidents and sometimes by design. "Cadets would look for a town or farm house, land, take a wrench and break a [spark]plug, this meant a couple of days of high living and good eats, in some towns drinks." Ground crews retrieved the pilots and machines. Nor did the leathernecks have much good to say about Army flyers. According to one leatherneck, "Two new army squadrons arrived. The majority have never seen a plane, and included draftees, cowboys, and a guard from the death house at Sing Sing."[21]

Life in the South meant becoming familiar with local wildlife. As temperatures rose, all sorts of things that crept and crawled appeared. Men collected snakes from tool drawers in the hangars. They went gator hunting in the afternoon. Swarms of mosquitoes clustering on window screens darkened barracks' interiors. Going outside often proved unbearable. In late February, Kelly complained, "The mosquitoes are getting to be awfully pest." The entire complement underwent inoculation for typhus. They did not enjoy it. One reported, "The whole company went to work this morning with their right arms in neutral and wearing a very sad expression. I do hope no more diseases are discovered before the end of the war."[22] A case of measles sparked rumors of quarantine. The following month, medical staff declared Lake Charles out of bounds for ten days due to outbreaks of meningitis and yellow fever. They ordered everything fumigated, even the mail.[23]

Sunday evenings brought welcome liberty and men often journeyed the eighteen miles to Lake Charles, a "fast town," where everything was "wide open, even burlesque shows." The pretty residential neighborhoods, "all lined with palm trees, the houses very large and southern," provided an attractive backdrop.[24] One weekend, the camp emptied as leathernecks went to town to spend their winnings from craps games. Those who remained behind used the barracks' open floor space to play "train" with rows of chairs, calling it the "Yaphank Special." New phonograph records caused "riotous rejoicing," with the old Victrola spinning till taps, followed by "Elmer's Dream" to "soothe us to sleep." At reveille, the boisterous strains of "Over There" filled the air. Toward the end of their stay in Louisiana, a pair of Marines rented "a decrepit old Dodge" and driver for $15 and undertook a cross-country jaunt to Beaumont, Texas, and beyond. One participant noted, "I spent $18 but I sure had my money's worth."[25]

Back in Philadelphia, Roy Geiger's small Aeronautical Detachment received fresh orders February 4, 1918, to proceed to NAS Miami. Cunningham raised the issue of the Marines acquiring their own field in a letter to Marine Commandant Barnett on February 6, suggesting the Corps secure "the old Curtiss Flying Field about five miles from Miami." A mild climate would obviate the need to construct permanent quarters.[26] Geiger's eleven officers and forty-one men entrained for sunnier climes the following day. After a short stay with the Navy at Coconut Grove, they moved five miles inland to the Curtiss field. Adjacent to the Everglades, it eventually accommodated the entire First Marine Aviation Force. Officers lived in apartments in Miami, enlisted men in tents at the field. Geiger quickly incorporated the Curtiss operation into the Marine program (renamed Marine Flying Field), commissioning civilian instructors and requisitioning school machines.[27]

Additional aircraft obtained through Cunningham's efforts allowed training to proceed, though not without difficulties and delays. All flying equipment was supplied by the Army via the Navy, and though promises of quick action were made, the reality proved different. In late March, Cunningham reported, "having trouble getting machines loose from the Army," adding a few days later, "If you have ever tried to do business with men like Admiral Benson and the Chief of Staff of the Army, and men of their type, you don't know how near they can come to driving a man crazy with their useless delays, etc." Of three Jennys taken over from Curtiss, one was wrecked by the end of the month. Congested freight lines only added to delays. Two aircraft received April 9 had been shipped February 18. As late as early May, only eight machines operated at the field.[28]

Efforts to obtain war-fighting aircraft for more advanced training also proved frustrating. At first expecting to be equipped with American-made Bristol fighters, the Marines soon learned "the Bristol is to be discontinued and DH [DeHavilland] substituted. I'm afraid," an officer at headquarters informed Cunningham, "We will have difficulty in getting delivery of these planes unless you come up here [Washington] to look out for it. Nobody seems to know what arrangements have really been made. We need to acquire some DHs on this side to test them. It would be very foolish to take a lot of these planes over to the other side without having some definite knowledge of whether they are any good or not, or would do the work expected of them." Ultimately, a tiny handful of DH-4s were obtained, but not in sufficient numbers and only after endless delays. In early June Cunningham announced, "4 DHs have been lost in shuffle." Even those that eventually arrived in Miami came with a warning: they should not to be flown except by the most expert pilots. "It would be a calamity to bust one up at present time."[29]

Other crucial supplies and equipment proved equally elusive. When Lt. Douglas Roben in Washington tried to obtain machine guns for fighter aircraft expected by the Marines, he hesitated, not wanting to "stir up the matter of armament in any discussion with the Signal Corps just now, for fear of getting in the way of the delivery of the machines [aircraft] themselves." An attempt to secure motor transport for the Miami field led to a flare-up at Navy headquarters. According to Lt. Harvey Mims, then handling aviation matters in Washington, "I finally quit dickering with [Albert C.] Read and [Robert H.] Gamble and took the matter up with Capt. Irwin direct and he has written to Yards and Docks."[30]

Clashes of personalities and inter-service rivalry did not help. When the Marines attempted to secure training slots at Air Service facilities, a drama ensued. According to Mims, one general had signed a letter to the Army requesting four officers and twenty enlisted men be instructed weekly at Wilbur Wright Field. Director of Naval Aviation Irwin deflected the Marines' request. Mims then made daily visits to Irwin's offices and had the time of his life, arguing with "that damned insect ensign [Gamble] of a pimp that you have wanted to knock in the face." Gamble said the Marines had it over the Navy regarding moving pictures. Mims balled him out in front of Putty Read, who said the ensign did not have rank to criticize Commander Towers (Irwin's principal assistant). Ens. Alan J. Lowry agreed. Mims concluded, "I would like to knock his block off and I think I could do it in about five minutes."[31]

The Miami field occupied a site about five miles inland on land reclaimed from the Everglades, now sandy black soil bordered by a canal extending fifty miles back to Lake Okeechobee. Captain Geiger, the camp commander, reported in early May that flyers faced difficulties maneuvering, taking off, and landing due to numerous soft spots in the field, while growing grass would be expensive and not yield results for one year. Student pilot Colgate Darden, a future congressman, governor of Virginia, and president of the University of Virginia, called it "the most unbelievable field that I have ever seen . . . was the field we trained on in Miami. The dust there, the sand was six or eight inches deep. The planes would struggle terribly in taking off. You would open them full throttle and they would struggle mightily at just a few miles per hour until they could extricate themselves and get up on top of the sand and then coast along to take off."[32]

Being situated next to a vast swamp made for difficulties finding emergency landing spots. One day while flying inland, Sam Richards spotted an aircraft upturned in tall Panama grass, luckily with only a broken propeller. On another occasion, Ed Kelly walked six miles into the Everglades to recover a crippled machine. The pilot had already hiked out and then had to walk back. As Kelly told his mother, "The Everglades are no place for me." Sand meant flies and one Marine reported, "The Florida sand flies are now busy making life very miserable." Work crews used crushed coral from the canal bank to fill holes in the field, one claiming, "We do not kill ourselves working and have a good time." Looking across the canal they could see "Negro convicts doing same work on road."[33]

It was at this moment, just as training began in Florida, that the huge program inaugurated by authorization of the Northern Bombing Group collided with the Marines' limited manpower resources. Manning six day bombing squadrons (later reduced to four) would ultimately require hundreds of pilots and observers and more than a thousand enlisted mechanics and other personnel. Responsibility for securing necessary manpower fell largely on Cunningham's shoulders. He visited the Officer School at Quantico to gather recruits, as well as volunteers that one flying veteran called "strays that Cunningham . . . picked up. I don't know where he got them."[34] In Miami, Roy Geiger signed up a few of the civilian students enrolled at the Curtiss Flying School by promising them commissions. On April 1, 1918, William McIlvain's group from Lake Charles, Louisiana, reached Florida, uniting the two Marine aviation units, but even the new combined force, plus additional recruits, fell far short of the numbers needed.

Cunningham quickly intensified his efforts. With the Navy's permission, he visited air stations at Pensacola and Key West where he signed up dozens of pilots who had completed basic flight training and wished to get into the action as quickly as possible, no matter what uniform they wore. According to Darden, Cunningham came to Pensacola and called together several aviators. He explained how the Marines were attempting to equip several squadrons of land planes for use in France. The Corps did not want to set up its own school because they did not have the time.

Instead, the Navy permitted transfers for as many as wanted. They would be sent to gunnery and advanced flying schools at Miami. A fair number decided to change uniforms. They had passed their flying tests but not received their commissions, which were slow coming from Washington. For some, the thought of piloting gigantic, cumbersome Curtiss H-16 patrol-bombers offered a good reason to transfer. The neophytes believed that if there were an accident, they would be killed by huge Liberty motors mounted above and behind the cockpit. As Darden recalled, "We decided we would depart for smaller and more manageable and faster planes." Even better, the Marines "gave absolute assurance we would be sent at once to France."[35]

Ultimately, 78 of the 135 Marine flyers who reached Europe were ex-Navy men. Pennsylvania native Sam Richards proved typical. He enlisted in the Navy at Washington on September 24, 1917. After Christmas, he reported to ground school at MIT, and following graduation moved on to Pensacola, where he soloed April 5, 1918. In June, Richards heard about the Marine option and decided to take advantage of it by transferring to the Corps. When he got to Miami Cunningham greeted the ex-Navy men "as if we were an answer to a maiden's prayers." Richards' transfer paperwork, however, was not completed until July 12, just as the Marines prepared to ship out for France.[36]

The Marines tapped all possible sources for recruits, but not every effort proceeded smoothly. One option was enrolling American veterans of Allied air forces. Several pilots in Europe who had flown for France joined naval aviation in the winter of 1918. One, Willis Haviland, formerly of the storied Lafayette Escadrille, served as Chief Pilot at NAS Dunkirk and would soon be on his way to Italy to assume command of the Navy's new station at Porto Corsini on the Adriatic coast. In April, Capt. Douglas Roben, then working at headquarters in Washington, became entangled in an effort to enlist Weston "Bert" Hall, also a veteran of the Lafayette Escadrille.

Hall wished to be commissioned as a Marine flyer and headed to Washington. Cunningham wrote a letter recommending he be enrolled in the Marine Corps Reserve as a captain. Roben took Cunningham's letter and approached Capt. Bernard Smith, a senior Marine aviator working in the offices of naval aviation, to get his signature. Smith reacted violently, saying Hall's conduct precluded acceptance. Roben wanted to go ahead and enroll Hall anyway, and would even serve under him rather than lose the veteran's services. Roben reported, "About the matter of Bert Hall, there is something very peculiar in Smith's attitude." He described the scene, "When he saw the letter he flushed and choked and was mad right away. He grabbed the letter out of my hand and tore it up."

Roben decided to let the matter rest temporarily, as "there may be reasons why he is not extremely desirable from a personal point of view. . . . Of course, if his appointment is going to bring about a breach with Towers, it wouldn't be a good thing for us." Roben blamed Smith and still pushed for Hall's enrollment. Perhaps Cunningham could see Towers directly.[37] In fact, Bert Hall never joined the Marines. And probably a good thing, as he was a thorough scoundrel. An original member of the Lafayette Escadrille, Hall soon earned the disdain of his comrades for his abrasive personality and slippery relationship with the truth. They eventually asked him to resign. A vigorous self-promoter, Hall became a mercenary and con man in the postwar years, spending part of that time in federal prison. The idea of him and the rather puritanical Cunningham being in the same Marine Corps defies credulity.

Despite all efforts, shortages of pilots continued, mandating some difficult decisions. In early May, Cunningham announced "new plans." Students showing the most aptitude would continue instruction as pilots, while those with less talent would serve as gunners. They should spend two hours per day training with their weapons. He warned, however, "The plan should be kept secret." Four days later he repeated the warning, "Don't spread word of the plan to use pilots as gunners, it might discourage some of the accidents?" In June, Cunningham admitted, "We will not get sufficient pilots from the Navy." He then ordered Roy Geiger to take pilots the Marines initially refused or disenrolled and use them as gunners, regardless of their flying ability. Other stopgap measures included using non-flyers in the rear seat. Karl Day chose Cpl. Frank Smith for one of the slots. Some objected, saying he was not a pilot. Day responded, "I said don't give a damn, he's on [the] Marine rifle team, he can shoot."[38]

The aviation program at Miami had many facets, evolving as the nature of the Marines' mission and complement changed. It included seaplane instruction at NAS Miami so that pilots could secure naval aviator designation, followed by land plane training, formation flying, acrobatics, rudimentary aerial tactics, gunnery, bombing, and reconnaissance skills.[39] Enlisted men received instruction as mechanics, armorers, and ground crew. According to Ralph Talbot, the day began at 4:45 a.m. "when I have to get up to take the bus to the field," and continued until dark, "generally about seven o'clock around here." Limited aircraft availability required certain ingenuity, including creation of two different instruction groups, one flying from daybreak until noon and a second from noon until dark, to get maximum use of the equipment.[40]

Aircraft on hand included "stick" Jennys with OX, OXX, and Hisso (Hispano-Suiza) engines, Thomas-Morse scouts, and eventually a few early-model DH-4s. Colgate Darden loved the Hisso Jenny, used for advanced training, said it had a beautiful motor. The men practiced acrobatics and dog fighting with the Thomas-Morse, but only made left-handed stunts because of the extreme torque generated by its rotary engine. The planes had no throttle, just a cut-off button, and pilots returned to the field covered in castor oil exhaust. The little "Tommy Morse" exhibited a disconcerting characteristic. When it reached a certain speed, according to Darden, "its tail would wiggle vigorously, and that scared us to death, because we felt it was only a matter of time before the tail would wiggle off. It never did . . . but when you put it into a little dive, you just had to be prepared for the wiggling at your back which was quite disturbing. . . . We never went out over water during Florida training. We'd fly out to the coastline in a minute or so, but we never went out over the Atlantic with land planes to speak of."[41]

Students simulated aerial combat with camera guns, one recalling, "We did a lot of practice with photo machine guns." As another explained, "You simply pressed the trigger and it clicked away. The development of that film would show us how accurate the firing was." Later they found out that the vibration of firing a real machine gun was very different. There were no towed targets or live air-to-air shooting. Bombing practice took place back in the Everglades, the targets being bales of hay covered with canvas marked with a red circle. Practice bombs consisted of sandbags attached to hooks beneath the wings.[42]

In the final weeks before the Marines packed up and headed overseas, a few DH-4 aircraft similar to those they would fly in France finally reached

Florida. This allowed at least some familiarization with the machine's handling characteristics.[43] Overall, new pilots received only modest training in acrobatics, bombing, and gunnery, and little of that on front-line warplanes. As to book learning and formal ground school, there was little or none. Karl Day later recalled, "We were never overly burdened with ground school subjects, in fact we didn't know much of anything except we were throttle jockeys. I taught some acrobatics there. We were short on theory, but I don't think that hurt us very much at that time."[44]

Accidents punctuated work at all training fields, and Miami was no exception. Inadequate equipment did not help. As late as June 21, only a few weeks before completing training, Harvey Mims in Washington passed some disturbing news along to Miami. He had spoken to Captain Irwin and been told the Army claimed the DH-4s sent to Florida after such long delays "were of an early type and not as reliable as a later type and should be handled with a little care." Writing about a mishap to Melville Sullivan, Naval Aviator No. 596, Cunningham in Washington observed, "Heard about accident to Sullivan. Operations peeved because they had not received immediate notice. Supposed it took place in Curtiss aircraft, underpowered with gunner standing in the rear seat and leaning over side with his gun, these machines should not be used for gunnery."[45]

Many decades later, Karl Day recounted a similar incident:

I did a silly thing, made up some gunnery ships out of wrecks, fuselage from one ship, wings from another, and "piled" [on] a scarff ring [machine-gun mount]. The thing was terribly out of line, but it was the only thing we could get at the time to give any machine gun training to our gunners; a terrific rivalry between these two shifts to see who'd get the most flying time. One of these gunnery ships was just sitting on the line not doing anything so I said I'd go to the gunnery field. This was across the canal, a little patch of sand. The gunners always had to crouch down in the landings and takeoffs because otherwise they would have deflected all the airflow from our tail surfaces, particularly the rudder. I saw I was going to overshoot, I began side-slipping to kill it, and here was my gunner, 6'2" standing as big as life looking over the side. There went my controls and we slipped into a spin at about fifty feet. Then we cracked up. The plane broke in two between me and the engine and between me and the gunner. The scarff ring went through the two rear cylinders of the engine, which was about the height of my neck. Only thing that saved him was that he hadn't buckled his seat belt. I got out with a few scratches, and gunner got out, he almost got hung before they got him loose.[46]

Some students got it right away. Others found the intricacies of flight daunting. Francis Mulcahy, acting as an instructor in Miami, had a trainee named Levy, nicknamed "the bounding Hebrew because he used to make some hard landings." He did not stay in aviation. Another pilot-cum-instructor, William Derbyshire, crashed on March 12 and broke both his legs. He was still recovering when the Marines headed overseas.[47]

Florida weather offered great challenges to flight operations, and life in general, especially as the season advanced. The head of the regional weather bureau claimed it would be plausible to fly in the summer, but by fall the field might be under water. The mosquitoes would be very bad, and hurricanes hit every few years. Atmospheric conditions would be very rough in the midday heat. Karl Day verified those predictions, writing home in early May, "Doing fair bit of flying these days, air here is rotten for flying, full of holes, bumpy and all that."[48] Just days before three squadrons departed Miami on their long journey to France, the Commandant received a description of conditions at the field. The men had been camped for months in tents on the edge of the Everglades without permanent accommodations. A Navy-Marine board of officers pronounced the field unfit for further use. Canvas hangars were ready to fall down. No grass could be grown on the sandy aviation field, which was also subject to flooding.[49]

As tourists already knew, Florida offered a range of recreational activities. Karl Day spent one Sunday at sea, departing at 6:30 a.m. and reaching the fishing grounds about ten miles offshore two hours later. His 30-foot launch "bounced around like a cork." With 150-foot hand lines, he snagged a 10-pound barracuda, "pronounced if spelled barrakewda," in less than five minutes, and later four small dolphins. A large sailfish broke the surface a short distance away. This only whetted Day's appetite for a three-day cruise to the Keys looking for tarpon, "the finest game fish in the world." Sam Richards claimed pilots occasionally went to the naval air station in Miami, got hold of an F-Boat, and flew to Bimini for liquor.[50]

Ed Kelly compared Miami to Asbury Park in New Jersey, "but with palms, everything costs double." The men could reach the beaches via a three-mile causeway across Biscayne Bay or on a ferry. Water temperature stood at 80 degrees, the surf was good, and Kelly practiced climbing coconut trees. On another occasion, he headed for the beach where he rented a bathing suit, "The colors are gorgeous, the louder the better. I never imagined I would appear in public in a canary yellow bathing suit, bordered in green." With access to automobiles, young men headed out in all directions. Kelly rented a "flivver" and motored up the coast to Palm Beach, but came away unimpressed. "Palm Beach is nothing," he opined, "the best it has to boast of is its exclusiveness, everything seems to be excluded. . . . The famous hotels were all closed."[51]

Back at the field, many swam in the canal that bordered the base. Alligators shared the water but did not seem to bother the bathers. Members of the party with rifles stood guard at each end of the canal. Occasionally a few Seminole Indians passed through the camp, "funny costumes, bare legs, short shirt to knees, head covering optional, anything from ribbons to a high hat. Motion pictures filled many evening hours."[52] Daytime monotony might be relieved by the twice-weekly gas mask drill. In early April, the station complement paraded in a Liberty Loan campaign. Kelly said, "It was my first parade and I felt as big as life with my squad. The people gave us a great deal of applause." While ants ruined a package of dates from home, Kelly called the local food excellent, especially the grapefruit, "all you can eat, my, what flavor."[53]

In early June, Captain Cunningham relinquished his post at Washington headquarters and headed south to take command of the squadrons training in Florida.[54] The camp soon began to buzz about rumored departure dates. As plans for European operations solidified, the expanded contingent underwent reorganization on June 16 into a headquarters detachment and four squadrons—A, B, C, and D, the commanding officers being Capt. Roy Geiger, Capt. William McIlvain, Capt. Douglas Roben, and Lt. Russell Presley, respectively.[55] Squadron commanders and their intelligence officers set off for France to familiarize themselves with the situation, select possible airfields, and establish liaison with Northern Bombing Group personnel.[56]

After several weeks' further training, the newly named First Marine Aviation Force—Squadrons A, B, C, and Headquarters Company—received long awaited orders to depart Miami and begin the journey "Over There."[57] Karl Day told his mother that everyone was in high spirits, "and in a couple of months you should hear about the terror spread among the Hun flyers by the Marines."[58] The entire contingent boarded the train July 13, passed through Philadelphia, and sailed from New York on board transport *DeKalb* on July 18, 107 officers and 654 men.[59] Facing an ocean crossing, the men proved to be brave, enthusiastic, patriotic, and thoroughly unprepared. One aviator recalled, "We had flown nothing but Jennies. We got one DH-4 and all of us in Miami got one flight in the first DH-4. . . . Our gunnery training had consisted of getting into the rear seat and using a Lewis gun, shooting at targets on the ground. None of us had ever fired a fixed gun in our lives. None of us had ever dropped a bomb in our lives."[60]

On board *DeKalb*, the former German raider *Prinz Eitel Friedrich*, the men went to their berths in the hold, the bunks stacked four high, eighteen inches apart. The ship departed the Hoboken docks at 12:30 p.m. and on the way out of the harbor passed a blimp and flying boat operating out of NAS Rockaway. The transport soon picked up its convoy, escorted by a cruiser and six destroyers. Once at sea, many "passengers" became violently ill and the hold "stunk of vomit." A lucky few managed to string hammocks on deck in the crew quarters. Commissioned personnel took turns with Officer of the Day duty and performed two-hour submarine watch every other day.[61]

The days that followed fell into a certain routine—breakfast, mad house drill, so named because of the men's chant "It's a mad house," trips to the ship's canteen to buy everything from crackers to plum pudding, setting up exercises, and performing candlestick watch in the main top searching for submarines. The men filled many of their off-duty hours playing Red Dog, a betting card name also known as high card pool.[62] In the evenings, there were movies and then "a scramble to see who sleeps on deck."

Everyone remained on high alert for U-boats. Around midday July 21 the lead cruiser fired at a dark object supposedly resembling a submarine conning tower, which disappeared at the same time general quarters sounded. Two days later, another general quarters rang, but turned out to be a false alarm. "Somebody leaned against the bell." On July 24, a suspicious black object off the port bow caused the gun crews on the transport to commence firing, soon joined by the cruiser and destroyers. Diarist John Benson recorded, "Don't know what it was, some shots close before it disappeared." The following day, "More target practice at boxlike object, don't know what we are shooting at but still we shoot for a better sport." One day turned into another, "Repetition of day before and day before that." For entertainment part of the ship's crew depicted the life and adventures of "Waco" Stewart from Texas, "very humorous and VERY burlesque. Another night movies with a few shots of 'The Flying Marines.'"

The would-be defenders of democracy entered the war zone July 28 and received orders to don life preservers. Nine destroyers from Brest joined the convoy, with the original escorts reversing course and heading home. Marine officers also received authorization to wear their new Sam Browne belts. As the ship neared France, more destroyers joined up. On July 30, a blimp appeared, replaced at noon by four flying boats that remained with the ship until it entered the Brest channel. Lookouts spotted land at 3 p.m. The men lining the rails saw many fishing smacks with their distinctive Breton red sails, as well as numerous forts along the shore.

From shipboard, the city seemed to be nothing but a vast collection of seawalls, chimneys, and castles. The men removed life preservers and emptied their canteens. They began transferring baggage to lighters and stayed up late reading French reprints of American newspapers. A few officers went ashore, heading straight for the Banque de France to get money and exchange government checks. Benson walked up and down the main street looking for things to do. No drinks to be had until six thirty so he bought raisins and fed his compatriots until time to return to the ship.[63]

According to Cunningham, the Marines reached France safely on July 30, "after a very pleasant trip but found upon arrival that no one knew we were coming or where we were supposed to go." He presumed the Aviation Division in the Office of the Chief of Naval Operations in Washington had failed to inform Captain Cone in Paris. Cunningham then called on Rear Adm. Henry B. Wilson Jr. Navy commander in France, who contacted Paris aviation headquarters. In the meantime, personnel remained on board ship for two days, until the vessel began preparations to return to the United States. One plan envisioned diverting personnel to Pauillac. Cunningham also claimed, "The Army tried to take charge of us but I refused." Instead, the men tarried for a short time at an Army rest camp just outside Brest. An attempt to retrieve the group's motor vehicles from Pauillac near Bordeaux revealed that "all our . . . trucks . . . had gotten mixed up and gotten into the Army pool." After untangling the snafu, the vehicles began the lengthy journey northward.[64]

The First Marine Aviation Force departed Brest for northern France in the early days of August. Corps histories claim no transportation had been arranged and that Cunningham personally requisitioned a train to move the men 400 miles to prospective bases near Calais. They boarded the famous French 40 and 8 railcars on a trip that lasted three days and two nights. Rations consisted of a loaf of bread and a gallon of canned peaches per man. Efforts to secure additional food at stations along the way ran afoul of the Marines' inability to speak French and local citizens' failure to understand English. After reaching Calais, the men were shunted to a British rest camp and treated to tea, black bread, and marmalade. After a short stay, they moved to unfinished aerodromes at Oye and La Fresne. Cunningham located his headquarters near Bois-en-Ardres on August 22.

Americans found much to say about their new home. Karl Day called France "picturesque and historic, but shy on baths." He described the aggressive behavior of street urchins. "On landing," Day noted, "the younger set of the inhabitants became afflicted with the gimmes, that is, gimme penny, gimme cigarettes. I guess they all began smoking at the tender age of four . . . shoes or sabots were almost entirely wood and made quite a noise on the cobbles." Thanks to the lengthy ride from Brest to Calais, Day had seen quite a bit of France, "not in Pullmans, but in cars called *chevaux* 8, *hommes* 40 . . . of course there were no chevaux in the car with us. The cars look like the toys you used to buy me at Asbury Park."

Settled in camp, he described his surroundings as a "very beautiful section of France, seems much like Long Island, rains frequently, making a most sticky mud." With liberty every night, Day visited the villages and wine shops, beer priced at four cents and a bottle of champagne at eight francs (about $1.40). Warned by a Belgian soldier that he would not sleep on clear nights, the young lieutenant did not understand until an air raid began at midnight. Anti-aircraft artillery and searchlights pierced the sky. Aircraft hummed overhead. When shells "burst [they] look like large stars."[65]

Acquiring warplanes to conduct operations proved a major challenge, one so severe squadrons would not mount their first unit raids until mid-October. Plans to secure DH-4s from the United States fell afoul of the many delays that bedeviled production, shipment, and assembly of aircraft from the United States to Europe. The first American-made DH-4 did not reach northern France until September 7.[66] Instead, on behalf of the Marines, the Navy negotiated an agreement with the British whereby Americans would provide three Liberty motors for one DH-9A.[67] Ultimately, the Marines obtained thirty-six aircraft, nearly evenly split between DH-4s and DH-9As.[68] Lacking warplanes and woefully deficient in adequate training, pilots, observers, ground crews, and other enlisted personnel spent the rest of the summer engaged in two critical activities: readying their aerodromes and temporarily joining RAF bombing units, much as their Navy partners were doing, though in far larger numbers.

While service with the RAF provided needed support to a war-weary ally and yielded invaluable lessons learned from hard experience, no independent Marine operations could take place without operating aerodromes. Oye aerodrome lay between Calais and Dunkirk, while La Fresne stood about seven miles southwest of Calais, both within easy flying distance of the German submarine facilities at Bruges, Ostend, and Zeebrugge. The bulk of personnel turned quickly to constructing barracks, hangars, mess halls, machine shops, offices, infirmaries, and all the other structures necessary to house the squadrons' activities. Other work parties labored under the summer sun draining and leveling flying fields, readying them to receive aircraft when they became available. In those first days, the men relied on motor transport loaned by the British. Many officers boarded with French families. After shelter arrived, they moved to the airfield, pitched their tents, and dug trenches around their new canvas homes to drain rainwater. "Perfection" oil stoves manufactured by Standard Oil Company provided heat. One flyer took a bath in an old jam tin.[69]

The aerodrome at La Fresne consisted of several farm fields, many of them growing crops of sugar beets. An Osage Orange hedge bordered one side and also enclosed farmers' buildings. The Marines acquired trucks and a bulldozer and ripped out the crop, leveled the field, and rolled it flat. The site was crowded, but workable. The only things lacking were the airplanes.[70]

Marine Corps aviation entered World War I with just a tiny handful of pilots and trained personnel and no assigned mission. By August 1918 they had assembled a concentrated aviation force on French soil, had settled into their new aerodromes, and were itching to join the fight.

Chapter Fifteen

Learning from the British

July–November 1918

An important part of negotiations concerning establishment of the Northern Bombing Group covered training and temporary service of Navy crews at RAF schools and with front-line squadrons. This coincided with the continuing need of British units for reinforcements. In mid-June, General Charles Lambe of the RAF announced he was ready to take as many Navy pilots for training as could be furnished him. Pilots were now being sent as rapidly as possible. Ten days later the first aviators departed the Air Service training facility at Clermont-Ferrand for duty with the Northern Bombing Group and assignment to the RAF.[1]

The Navy eagerly placed scores of pilots and observers at the disposal of the British in northeast France. This group included aviators assigned to day and night bombing squadrons flying DH-4, DH-9, and Handley Page aircraft. It also counted a few piloting deadly Sopwith Camel scouts, including the Navy's only "Ace" of the war, David Ingalls. Over a four-month period, they participated in day and night assaults against a wide range of targets. A postwar analysis concluded, "The training the personnel received during this period proved exceptionally valuable."[2]

Navy day bombing crews trained at Clermont-Ferrand began service with 218 Squadron in mid-July.[3] Pilot-observer teams participated in at least three raids, though for some their time with the RAF lasted much longer, participating in up to a dozen missions. Marine Corps officers selected to command day bombing squadrons and dispatched to Europe ahead of their ground units also served with the British. The practice of placing American personnel with the RAF continued until the last few weeks of the war. In fact, Northern Bombing Group personnel flew far more missions with British squadrons than they did with their own.

No. 218 Squadron came into existence as a day bombing unit on April 24, 1918, in Dover, England, commanded by Canadian maj. Bert Sterling Wemp, DFC. They crossed over the Channel to Petite-Synthe near Dunkirk several weeks later and joined No. 82 Wing, 5th Group, RAF. The squadron carried out its first raid in early June.[4] Composed almost entirely of new pilots, only Wemp and fellow Canadian flight commanders John Chisolm and William Cleghorn had any experience at the front. Crews included a wide spectrum of personnel drawn from several of Britain's overseas dominions. In July, the squadron relocated to the aerodrome at Fréthun, about four miles southwest of Calais as a defense against heavy German aerial assaults in the Dunkirk area. Between early June and the Armistice, 218 Squadron conducted 117 raids, dropped 94 tons of bombs, and shot down 38 enemy aircraft. The unit flew the Airco DH-9 capable of a top speed of barely 85 mph, and only 75 mph or less when loaded. Monthly wastage of this unsatisfactory aircraft approached 70 percent. Nonetheless, several American pilots liked the way the machine handled. David "Eddie" Judd, a Harvard graduate and veteran of the Lafayette Flying Corps, called it "a fine little sewing machine."[5]

Among the first Americans to reach 218 Squadron were David Ingalls, Ken MacLeish, and Eddie Judd, along with their observers Randall Browne, Irving Sheely, and Sidney Huey. They had been sent ahead of other Navy crews with the specific intention of readying them to become flight commanders in the first day bombing squadron. No sooner had they arrived than they were thrown into battle, without even a familiarization flight on their new DH-9s, as part of a raid targeting the Zeebrugge mole and warships sheltered there. The three-hour mission resulted in six damaged aircraft and one more that failed to return.

Caught in the heavy barrage, MacLeish and observer Irving Sheely barely escaped with their lives. MacLeish observed, "Archie [anti-aircraft] was almost ten times as noisy and friendly as I ever hope to see him again." A shell exploded just beneath the tail of their aircraft, punching a hole in the rudder, rocking the machine, and sending it into "all kinds of dives and gizzy-wiggles before I could get control again." No sooner did MacLeish stabilize the aircraft, than the motor quit, sputtered, and gurgled, and then caught on five cylinders. Fighting a strong head wind, the Americans barely maintained flying speed, but managed to limp across the lines. After three raids, MacLeish, Ingalls, and Judd departed the squadron, assigned other duties preparing Northern Bombing Group personnel and equipment for deployment in the field.[6]

In late July, pilots George Moseley, Chet Bassett, Albert Johnson, and Charles Beach replaced the initial Navy contingent. In the weeks that followed, they made as many as a dozen flights, attacking the same complex of coastal bases ranging from Ostend to Zeebrugge. Moseley described the routine in a letter home. "The squadron goes over the lines every day, sometimes twice, weather permitting, and bombs the big ammunition dumps, the factories, railroads, troops, etc. . . . It is very interesting work as we have to go through a very heavy fire from the anti-aircraft guns and are harassed by German chasse machines most of the way to our objective and back." Nonetheless, Moseley believed "this is much better work than water-chasse, because now I feel as though I were really accomplishing something."[7]

The squadron often carried out two missions a day, meaning five to six hours in the air. Although frequently bracketed by "Archie," Fred Beach recalled that they rarely met enemy fighter aircraft. "We always flew in close formation," he said, "and had a good chance of driving off an attack until some of our escort could come up." As to the actual bombing, "It was pretty crudely done . . . and the pilot had to drop them [bombs] besides keeping in formation, dodging the 'Archie,' taking care of his motor, and watching out for enemy planes." Concerning the day bombers he flew, Beach said, "Our machines were DH-9s with Puma motors of much too small horse power. They were fast on the level and handled beautifully, but climbed sluggishly and, when loaded, their ceiling was far too low for comfort."[8]

Moseley described the launch of an over-the-lines mission in detail. Ground crew started the engines at 7:30 in the morning, with the machines—16 aircraft, each with 8 small bombs affixed beneath the wings—already aligned in the positions they would maintain while in flight. At exactly the scheduled time, the squadron commander standing on the far side of the field raised his hand and the flight leader gunned his engine and began accelerating across the grass surface until he rose into the air. Successive aircraft, first on the left and then on the right, followed him aloft. Once airborne, pilots assumed their proper places and began climbing slowly to an altitude of 15,000 feet. Sopwith Camels flying escort soon joined them. The entire group headed out to sea and worked its way up the coast, beyond range of threatening anti-aircraft fire. This particular raid targeted the mole at Zeebrugge. All along the way, pilots and observers strained to detect German fighters, but avoided contact.

With the objective in sight, attacking aircraft turned toward shore and commenced their bombing runs, dodging the thickening curtain of "Archie" every yard of the way. A thin white mist provided a wisp of protection. Leaving Zeebrugge, the bombers pulled into tight formation with the escorting Camels and raced for home. Once across the lines, some of the aircraft broke ranks. Moseley attacked a target positioned just outside of Dunkirk, both he and his observer blasting away with their machine guns. Then they raced along the beach, roaring along just a few feet above the waves, all the way back to Calais. Safely landed at the aerodrome, Moseley counted four or five holes in his wings. An observer in another machine was hit by a piece of shrapnel that penetrated his flight suit and clothes, but did not injure him. The squadron mounted a second raid against the same target that very afternoon.[9]

Raids and attendant duties did not fill every hour of the day and the Americans found a variety of diversions to keep them entertained. Dinner with their respective messes offered lighthearted banter and good fellowship. Navy officers thoroughly enjoyed their British compatriots. Fred Beach described the RAF flyers as "youngsters from seventeen to twenty-two—nothing sluggish about them—a nervy, reliable bunch and always full of pep." Some Americans spent their spare time visiting comrades at NAS Dunkirk. David Ingalls and Eddie Judd secured a motorcycle and raced around Flanders. They also went to Dunkirk for a meal, Ingalls noting "we greatly enjoyed the food. The RAF grub is not for a luxury lover like myself."[10]

On another occasion, he commandeered a Cadillac motor car and visited old friends Ed "Shorty" Smith and Di Gates in Dunkirk. On the way back to Calais, Ingalls met the executive officer at St. Inglevert who had corralled some WAACs[11] and brought them to dinner at the chateau near the field. "They just tore around saying how 'priceless' and 'hopping' everything was." Robert Lovett, now commander of the Night Wing of the Northern Bombing Group, visited a few days later bearing words of reassurance. Mail, magazines, and packages from home—especially those containing candy and tobacco—also raised spirits. On July 18, Ingalls and two British officers visited an army remount station where they secured horses and enjoyed a fine cross-country ride. Unfortunately, the horse of one of the flyers slipped, fell on the man's leg, and crushed his foot. He was hauled to the hospital in an ambulance. Major Wemp immediately posted a notice forbidding further horse riding because it was "too dangerous."[12]

Another day bombing unit, 217 Squadron, also benefited greatly from the infusion of American personnel. The squadron traced its lineage back to the RNAS seaplane station established at Dunkirk in October 1914. It was designated 17 Squadron on January 14, 1918. Following consolidation of RFC and RNAS squadrons on April 1, 1918, the group became 217 Squadron, RAF. Flying DH-4 bombers, the unit attacked enemy airfields and bases in Belgium. The squadron operated out of Bergues aerodrome, just five miles south of Dunkirk.[13] Part of 61 Wing, other units there included 202 Squadron—a reconnaissance/photography formation equipped with DH-4s—and 213 Squadron flying Sopwith Camels. Given its location within the larger Dunkirk complex of bases and military facilities, the field endured frequent attacks. Writing from the safety of Clermont-Ferrand, Navy pilot Kenneth MacLeish reported, "We just heard today [June 18] that 40 Hun machines bombed D[unkirk] for three nights straight and dropped 700 bombs. They absolutely wiped out the H[andley] P[age]s, killed my old flight commander [John de C. Paynter] just as he was getting into bed, and killed several over at 217 Squadron where Di was."[14]

Of all the Navy pilots who flew with 217 Squadron, Archibald "Chip" McIlwaine compiled the longest record. A member of the First Yale Unit, he was one of the earliest Navy pilots dispatched to Europe. He received additional training at Hourtin and Moutchic and remained for a period as Chief Pilot. After a stint at Clermont-Ferrand, McIlwaine joined the pilots training for the Northern Bombing Group and was assigned to 217 Squadron, then flying out of the aerodrome at Crochte, about eight miles south of Dunkirk.[15]

McIlwaine and the rest of his flight attacked the mole at Zeebrugge on July 29. They came back the next day, navigating by compass through dense fog, and bombed four German destroyers from an altitude of only 1,700 feet, dodging anti-aircraft fire all the way. They returned to Zeebrugge yet again on July 31. Other raids included attacks on German supply depots from altitudes as low as 250 feet. Though many aviators found such a regimen physically and mentally exhausting, McIlwaine seemed to thrive on it. He noted how crews slept at the airfield the night before a raid, rising to a good breakfast of oatmeal, eggs, and coffee, "and the indispensable jam or marmalade." Pilots soon "sauntered" out to the field. "Then we sailed away. It was distinctly *de luxe.*" Flying, he noted, was a "test of nerves, a sport for which you had to be kept fit. Shave in the evening, take a hot bath, get your shoes shined whenever you liked—a perfect snap."[16] Recalling his time with the British, McIlwaine listed twenty-six flights over the lines, engaged in four dog fights, and strafed enemy trenches. He boasted, "In short, [I] had a devil of a good time."

American crews continued flying with 217 Squadron for several months, sometimes with tragic results. A bizarre accident occurred on August 20 that cost the lives of two Navy aviators, Ens. Thomas McKinnon (pilot) and Yeoman Marlon O'Gorman (observer). The pair set out in a DH-4 on a morning offshore anti-submarine patrol near Dunkirk. While flying straight and level their propeller flew off and fell into the sea. The aircraft continued on for about thirty seconds and then banked toward shore, stalled, fell into a spin, and crashed into the water. A nearby RAF pilot, Lieutenant R. G. Shaw, witnessed the accident, flew back to St. Pol, and sounded the alert. Destroyers sped to the crash site where they recovered O'Gorman's body, kept afloat by his inflated life belt. McKinnon was never found.[17]

Navy airmen also served with British heavy bomber forces, operating enormous Handley Page machines. Lovett and McDonnell paved the way, participating in a dozen night raids in late March and early April. During the summer, several additional crews compiled considerable time with 214 Squadron, stationed for much of the period at Coudekerque, not far from Dunkirk, or at St. Inglevert, southwest of Calais. The unit assembled first at Coudekerque on July 28, 1917, designated initially as 7A Squadron, RNAS. From the beginning, its role consisted of carrying out heavy, night bombing raids against targets in Belgium. On December 9, 1917, the group was renumbered 14 Squadron, and with the April 1, 1918, RAF consolidation, became 214 Squadron. After enduring a series of heavy air raids in June, the unit relocated to the aerodrome at St. Inglevert.[18]

Americans seconded to 214 Squadron flew the Handley Page bomber, mainstay of Britain's heavy bomber forces in the last year of the war. The machine was conceived early in the conflict following a meeting between Captain Murray Sueter, Director of the Admiralty's Air Department, and manufacturer Frederick Handley Page, who eventually earned the unofficial title, "Father of the Heavy Bomber." The Royal Navy desired "a bloody paralyzer of an aircraft." The eventual result, the Handley Page 0/100, first flew in December 1915. The initial squadron formed in August 1916 and carried out its inaugural raid in March 1917. A total of forty-six of these huge aircraft were produced. Deliveries of the improved 0/400 model commenced in early 1918 and continued for the remainder of the war. The new model incorporated a strengthened fuselage, greater bomb load, increased speed, and higher maximum altitude. A few machines carried massive 1,650-lb bombs in later raids. Over 400 were delivered to the RAF before the fighting ended.[19]

The first Navy pilots reported to 214 Squadron at Coudekerque June 23, 1918, Leslie Taber, James Nisbet, Jesse Easterwood, and Phillip Frothingham, all ensigns. Others quickly followed. As more personnel passed through the RAF gunnery school in Texas or pilot training at Stonehenge in England, the number of Americans increased substantially, amounting to seven pilots and forty enlisted men by August 10. Six officers went on bombing missions in September as aerial gunners. Lt. (jg) Alexander McCormick died one evening shortly after returning from a raid when he inadvertently walked into this aircraft's spinning starboard propeller. Many attacks during this period proceeded not against submarine facilities in Belgium, however, but against enemy troop and transportation targets, railroad yards, and junctions.[20]

By mid-October, largely American crews operated two of the huge night bombers, including Lt. (jg) Moseley Taylor, who had labored for a time at Stonehenge as an instructor.[21] As Capt. David Hanrahan reported in his summary progress report of November 11, 1918, a number of night pilots were in training with 214 Squadron until the Armistice. Two, Moseley Taylor, and Ens. William Gaston, were carried as regular first pilots, having Handley Pages assigned to them. They remained with the squadron up to the time of the cessation of hostilities. As a measure of how valuable the Americans' service had been, Hanrahan noted, "When Number 214 Squadron was transferred in the latter part of October, the Squadron Commander made every effort to retain our personnel."[22]

By any measure, allowing Navy and Marine flyers and ground crews to serve with front-line RAF units provided enormous benefits to both Great Britain and the United States. The flow of fresh manpower into beleaguered squadrons provided these stressed formations with necessary infusions of bravery, enthusiasm, and a willingness to learn just as the tide of battle turned. For the Americans, assignment to experienced operational units allowed untested reserve officers to acquire the skills necessary to make combat-ready aviation a reality. Whether flying by day or night, both parties received exactly what they greatly wanted and desperately needed.

Chapter Sixteen
Operating against the Enemy

The Campaign Begins

By the end of July 1918, the Northern Bombing Group seemed poised to begin independent operations. Captain Hanrahan had gathered much of his headquarters staff. Three complete squadrons of Marine aviators were on their way to new aerodromes in Flanders. The Navy took control of a large assembly and repair facility at Eastleigh, near Southampton. Arrangements with the RAF promised easy access to construction supplies, fuel, and maintenance in the field. American crews were ready to begin ferrying Caproni bombers from the factory on the outskirts of Milan across the Alps to France and the air bases that had been established in the Pas de Calais.

The first operational endeavors consisted of efforts to fly Caproni bombers across the Alps from northern Italy to aerodromes in Flanders. Nagged by intractable shipping shortages and railroad bottlenecks, this seemed the option most likely to move aircraft to the battlefield in the shortest time. It turned out to be an unmitigated disaster, however, exposing both the mechanical failures of the aircraft and inadequate training of flight crews.

Naval aviators prepared for their cross-continent mission by undertaking a series of less demanding flights shuttling bombers destined for the Italian air force from the factory to the flying field at Gioia delle Colle near Brindisi. The two-day journey included stops at Pisa, Centocelle, Capua, and Foggia, with Americans normally paired with enlisted Italian pilots. A dozen Navy flyers participated in the activity, but four aircraft failed to reach their final destination due to accidents. Broken propellers or malfunctioning engines caused others to abort their missions. One aircraft mistakenly headed north rather than south and landed just fifteen miles short of the Austrian lines.[1]

As soon as the first Capronis allocated to the American services were delivered in late July, efforts commenced to fly them across the rugged mountains. Flight operations on the Italy–Flanders route began badly and rapidly grew even worse. Mixed Army-Navy crews manned the initial batch of aircraft, with later efforts undertaken by exclusively Navy personnel. Numbered B-1, B-2, up to B-17, the first machine crashed at Turin. The second, with a mixed service crew, went down at Sens, France, on July 28. A bomber flown by Ensign Harry Krumm and Ensign Nesbit crashed at Turin-Mirafiori four days later. That very same day, Lt. (jg) Reginald Coombe's machine went down near the factory at Taliedo while he was making a landing during a test flight. Another bomber with an Army crew was completely destroyed at Bra.

In early August, Lieutenant McDonnell reached Italy after his stint at Navy headquarters in Washington and took control of the operation; during the next two weeks several more aircraft began the hazardous journey north. Most experienced repeated engine trouble. Several caught fire due to faulty carburetors. On August 17, Ens. Alan Nichols and Ens. Hugh Terres, as well as mechanic Orrin Hartle, died in a violent crash. One witness recalled, "The machine was broken into little pieces, utterly demolished, their bodies had been smashed." Other pilots encountered great difficulties crossing the Alps or making their way northward through France.[2] McDonnell's own machine crashed in the mountainous terrain near Lyon, catapulted upon impact, and "flew into kindling wood." Reginald Coombe's odyssey lasted more than two weeks. Of the seventeen aircraft that set out from Italy, only eight made it safely to France. In all, five crewmen died.[3]

Despite staggering difficulties encountered ferrying bombers from Italy to northern France, the arrival of the first aircraft at St. Inglevert sparked hope within the aircrew and ground personnel that the Northern Bombing Group would soon begin combat operations. Mechanics rigged bombing apparatus and navigation lights, installed additional instruments, rearranged landing gear wires, and readied machine guns. On the night of August 15, the newly configured aircraft B-5, fitted with an extra fuel tank, participated in a multiplane raid conducted by 214 Squadron against the Ostend docks, dropping 1,050 pounds of bombs before returning safely to base. In the next two weeks, two more sorties were attempted, but both times the machines turned back due to failure of the cranky Fiat engines that had been substituted for Isotta-Fraschini motors used on earlier machines. On the first mission, the pilot landed the plane with the full bomb load still on board. On the second attempt, the Caproni B-5 crashed upon landing, a total wreck.

Investigations conducted by mechanics in the field and technical experts in Paris and elsewhere identified many defects in the Fiat engines. These included badly machined and assembled parts, faulty carburetors, and fuel feed systems subject to frequent breakdown. Inability to resolve problems with the Fiat engines eventually caused headquarters to forbid use of Capronis for further combat missions. This left the Night Bombing Wing with two aerodromes and hundreds of pilots and ground personnel, but no aircraft with which to operate. In desperation, the Navy turned to the British, seeking to acquire Handley Page bombers in exchange for Liberty engines. An arrangement worked out between the U.S. Army and Navy and British authorities eventually allocated twenty bombers to the Americans, ten to each service. A few were undergoing tests when the war ended.[4]

Whatever mechanical failures bedeviled pilots and crews, nothing compared to the tragedy that unfolded at St. Inglevert on September 15, 1918. Pilots Phillip Frothingham and Clyde Palmer, assisted by crew member Chief Electrician Austin Underwood, took Caproni B-11 aloft for a short test flight. The same crew had only recently ferried the machine across the Alps from Italy, a process that consumed more than two weeks. According to the station log and a later statement by Night Wing commander Robert Lovett, at 3:43 p.m. the aircraft made a good landing at about 30 degrees into the wind. After taxiing no more than thirty feet, the nose of the Caproni dipped, striking the ground, causing the fuselage to fold up beneath the pilots' seats, pinning both men in place between the wreckage and the middle gas tank, with the weight of the machine upon them. A fire broke out and Underwood tried desperately to cut off the motors and extricate the pilots. Flames quickly engulfed the aircraft. Numerous members of the ground crew and casual onlookers rushed to the stricken plane but could not save the trapped aviators. After the fire cooled, their charred remains were identifiable only by the metal tags they wore.[5]

A court of inquiry at the field began taking testimony almost immediately in an attempt to discover what caused the disaster. According to survivor Underwood, the landing gear had sunk into the mud on the wet field and the plane "turned over on its nose." The propellers hit the ground and the motors began to race, causing a fire in the carburetor. "The machine instantly burst into flames." Fire reached the trapped pilots in only thirty seconds. It could not be determined if they remained conscious or had been choked prior to the conflagration. A second witness added the fact that the incident occurred just after the aircraft turned toward the hangars. Thirty men, a water tanker, and the station ambulance raced to the machine. Lieutenant Lovett jumped into his car and headed for the hangar to fill the vehicle with fire extinguishers and then dashed back to the burning airplane. The ad hoc firefighting crew exhausted twenty-five pyrene extinguishers with no effect. Chief Electrician Underwood had to be dragged from the aircraft, "as he stayed until the very last moment endeavoring to get the men out and shut off the machine."

Lovett had nothing but praise for his men's action during the fire. He reported that Underwood "displayed great bravery . . . although surrounded by flames succeeded in cutting off one motor and was endeavoring to cut off the other when he was pulled out by Lieutenant C. R. Johnson." Lovett related how several men grasped red hot steel struts attempting to pull the wings off the plane. "The entire emergency party displayed the utmost coolness and contempt of danger." The Board determined the pilots died "by means of the accidental capsizing and burning of Caproni aircraft B-11 causing fatal burns."

The cause of the accident remained uncertain, however. Lt. (jg) Moseley Taylor, a pilot with considerable experience operating large bombers, believed the Caproni's notoriously weak undercarriage might have given way during the turn, or that with the breeze blowing from the rear a puff of wind under the elevators pushed the nose down as the aircraft taxied. Lovett hypothesized that the axles might have been damaged during a sharp turn after landing. This caused the aircraft to move more slowly across soft, wet ground. The pilot then increased power to the motors just as the machine entered a small, chalky depression in the field, causing it to nose over. Crushing of the nacelle pulled all the motor control wires, causing the Fiat motors to race, leading to the carburetor fire and disaster.[6]

In addition to efforts aimed at conducting war missions and making aircraft serviceable, other major activities at the night bombing aerodromes involved construction and training, interrupted by visits of various dignitaries and liberty in nearby towns and cities. Additional Capronis arrived sporadically. Eventually eight machines made it to northern France, often after odysseys lasting three or four weeks. Reginald Coombe and Oliver P. "Ollie"

Kilmer brought Caproni B-16 to St. Inglevert from Paris on September 2. Coombe was a member of the First Yale Unit, having already served as Chief Pilot at NAS Le Croisic and received flight instruction at the Caproni school at Malpensa, Italy. Assistant pilot Kilmer, at thirty-nine one of the Navy's oldest flyers and a veteran of the Great White Fleet and the American intervention at Vera Cruz, received his Airman's Certificate as an enlisted man in January 1917 and was designated as a Naval Aviator on October 2 of that year.[7]

Once operations began, a parade of visitors journeyed to St. Inglevert to inspect the Navy's signature bombing program. Captain Hanrahan and Colonel Spenser Grey arrived at the beginning of September in an aircraft piloted by Grey. The two senior officers were extremely embarrassed when the plane nosed over while they were making a turn on the field after landing. Fortunately there were no injuries, except perhaps to their pride.[8]

Members of the House of Representatives Naval Affairs Committee on an extended tour in Europe arrived at St. Inglevert September 3, including Congressmen Frederick C. Hicks (New York), James C. Wilson (Texas), and William B. Oliver (Alabama).[9] The following month, Adm. Henry Mayo, commander in chief of Atlantic Fleet, with a retinue that included Naval Attaché Richard Jackson, aide Capt. Ernest J. King, and Cdr. John V. Babcock of Admiral Sims' staff, stopped at the field during his lengthy European fact-finding mission. Two weeks later, Hanrahan accompanied Capt. Noble Irwin, the director of Naval Aviation, during the latter's first and only visit to the war zone.[10]

Throughout the fall, parties of enlisted personnel performed varied duties with neighboring 214 Squadron, both learning and assisting, as on October 24 when eight quartermasters, electricians, and mechanics moved across the aerodrome to adjacent British facilities. The following week, a large detachment of three officers and fifty-four enlisted men traveled to Calais to attend the funeral of Maj. Douglas Roben, USMC, commander of Day Squadron 9, who died on October 27 from complications of influenza. Anticipating the war might last into 1919 and hoping for the arrival of Handley Page bombers or re-engined Capronis, Lieutenant Lovett and several others set out on a late October foray toward the front lines to identify possible new aerodrome sites.[11]

Over at Campagne, the aerodrome designated to accommodate Night Bombing Squadrons 3 and 4, work advanced more slowly. Construction proceeded on hangars, barracks, provisions' lockers, shops, a YMCA hut, and the sick bay. Bombs arrived, but no airplanes were ever received or

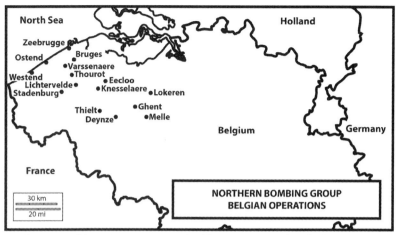

Northern Bombing Group personnel either serving with Allied air units or conducting their own independent raids attacked a wide range of targets in Belgium during the closing weeks of the war.

flight operations conducted. Bluejackets stationed at Campagne often served as a sort of labor pool for other projects. Every day, large work parties commuted to the nearby RAF depot at Guines to construct accommodations for several hundred anticipated American personnel. Workers traveled to the Marine field at La Fresne as well. Captain Hanrahan visited in early October and returned a few weeks later with Captain Irwin. Influenza struck in mid-October, temporarily paralyzing operations. When a new draft of forty-seven men arrived from Pauillac, they were quarantined due to an outbreak of meningitis. On October 28, a working party began dismantling the hangars they had just erected, preparatory to relocating to a new site closer to the battle front. At the same time, recognizing the unlikelihood of mounting night bombing raids in the foreseeable future, London headquarters requested that twenty night bombing pilots previously qualified to operate H-16 flying boats be detached for temporary duty at NAS Killingholme.[12]

Even as the Navy shuttled small groups of pilots and observers to and from RAF units, there remained a substantial number of surplus pilots who had been trained at Clermont-Ferrand and elsewhere to carry out duties now assigned to the Marine Corps. This resulted in several flyers being attached to French and British squadrons in the Dunkirk area. Their exploits constituted the Navy's most extensive involvement in one-on-one aerial combat during the entire war. David Ingalls and Kenneth MacLeish fought with 213 Squadron, a unit equipped with Sopwith Camels, and with whom they had served in April. During his time at the front in August and September, Ingalls scored the aerial victories that made him the Navy's first and only "Ace" of World War I.

After pestering everyone he knew for reassignment from aerodrome construction duties at Oye, David Ingalls joined 213 Squadron at Bergues on August 9 and completed a short practice flight at the controls of a Sopwith Camel within moments of his arrival. By 4:30 p.m. he was off on a high offensive patrol across the lines with six other squadron pilots. Ingalls scored his first victory August 11 when he and Captain Colin Brown, RAF, combined to down an enemy Albatros two-seater. Two days later, he participated in a massive aerial assault involving both British and American squadrons against the German aerodrome at Varssenaere, Belgium.[13]

In the weeks that followed, Ingalls scored at least five more victories, including destruction of a kite balloon near La Barriere. He was also named an Acting Flight Commander. On September 15, he led Flight "C" as part of a strike against Uytkerke aerodrome near Zeebrugge. Commanded by Captain C. R. Swanson, RAF, four flights from 213 Squadron unleashed 80 bombs and expended 2,200 rounds of machine-gun ammunition, inflicting extensive damage and casualties. On the return flight, Ingalls shot down a Rumpler C observation machine in flames near Ostend.

For this exploit and his other work with the unit, squadron commander Maj. Ronald Graham recommended Ingalls for the Distinguished Flying Cross, and offered high praise, saying, "His keenness and utter disregard of danger are exceptional and an example to all. He is one of the finest men the Squadron has ever had." The actual award citation noted, "Alone and in conjunction with other pilots, he shot down at least four enemy aeroplanes and one or more enemy balloons." After eight hectic weeks with 213 Squadron that included continuous patrols, raids, and strafing attacks, Ingalls departed Flanders in early October to assume a position as head of the Flight Department at Eastleigh. His responsibilities there included final preparation and test flying of Liberty DH-4 and DH-9A aircraft assigned to Marine day bombing squadrons in France.[14]

Relinquishing that same post was Kenneth MacLeish, Ingalls' longtime training partner and squadron mate. Now it was MacLeish's turn to join 213 Squadron, with whom he had flown back in April. Having just been through two months of intense combat, Ingalls could not understand why MacLeish seemed so eager to go, telling his mother, "Ken to go out with 213 Squadron. Imagine a fellow engaged to a wonderful girl, as Ken is, wanting to go looking for trouble." With wisdom borne of hard experience, the nineteen-year-old observed, "Also I might be able to give him some good dope, good pilot as he is. It's the first scraps that are a bit dangerous and maybe I could ease him over a few bad points."[15]

When MacLeish replaced Ingalls at 213 Squadron, he began his assignment auspiciously, downing a Fokker D.VII October 14 near Thourout on his first cross-lines sortie.[16] That same afternoon, 15 Camels initiated another patrol along the Channel coast. Near Dixmude, they spotted 14 Fokkers at 8,000 and 12,000 feet. A melee erupted, with several German and three British aircraft lost. According to the station log book, "MacLeish was last seen attacking about seven Fokkers single-handedly." Sadly just days before he informed fiancée Priscilla Murdock that he had been to London to request transfer back to the United States as an instructor as soon as his time at the front was over. His body was not discovered until the day after Christmas.[17]

One other group of Navy pilots under Northern Bombing Group jurisdiction also joined an Allied unit in the final weeks of the war. By summer's end, operations at NAS Dunkirk had come under the aegis of the Northern Bombing Group. At the same time, dreadful weather conditions and inadequate equipment greatly limited the number of patrol missions executed, sometimes canceling all flight activity for days at a time. In response, several aviators, including those trained at Clermont-Ferrand during June and originally slated for assignment to proposed Navy bombing squadrons, lobbied to be allowed to fly with the Allies. In late September, they got their wish, posted to temporary duty with Escadrille de Saint-Pol flying SPADs and operating out of the St. Pol-sur-Mer aerodrome just southwest of the city.

After a few practice flights, the Americans, including Di Gates, commanding officer at NAS Dunkirk, George Moseley, William Van Fleet, and Charles Beach, began patrols across the lines, attacking retreating enemy troops, even dropping emergency supplies on isolated French soldiers. On October 4, these four participated in a seven-plane patrol led by Captain Delasalle over the Ypres salient. During a dust-up with a pack of German fighters, Di Gates and his aircraft disappeared. George Moseley reported, "We were looking for Huns in the Ypres salient. . . . Art [Gates] did not come back. He was with us when the attack began, after that nobody saw him." In fact, Gates and his machine made it safely to earth behind German lines where he was soon captured and held as a prisoner, first in Belgium and then in Germany. Twice he tried to escape and was recaptured both times, the second attempt taking him within steps of the Swiss border. Gates was freed in late November and returned safely to the United States in early February.[18]

Chapter Seventeen
Operating against the Enemy

Day Bombing

With efforts to launch punishing nighttime raids stymied by shortages of equipment, attempts to implement the complementary day bombing campaign inched ahead. Initial plans for supplying aircraft to the Northern Bombing Group called for shipping American-manufactured DH-4 day bombers to the Navy's enormous logistical facility at Pauillac, on the banks of the Gironde River in southern France, about thirty miles downstream from Bordeaux. A large assembly facility was constructed and a flying field laid out. The completed machines would be tested here and then ferried by air over four hundred miles to aerodromes in Flanders. The first aircraft from the United States were assembled and flown in August.

This arrangement encountered many obstacles. Aircraft and engines were slow to arrive from the United States. Cargoes and necessary parts were diverted to other ports. Overcrowding at the riverfront docks and endless paperwork slowed the vital processes of unloading and then assembling the equipment. Most engines required rebuilding due to poor workmanship and inspection. Aircraft required extensive modification. Even the smallest task sometimes entailed frustrating delays to those trying to expedite the process. Moving airplanes many hundreds of miles to front-line aerodromes presented additional difficult challenges.

Kenneth MacLeish, detailed to Pauillac as chief pilot and charged with final test flying of airplanes, documented the frustrations experienced by many. He noted on August 6, "Not a blessed thing done today. The people unloaded a few crates at ten o'clock this morning, and we didn't see them again all day—our hangar is two hundred yards from the dock. They couldn't make two hundred yards in seven hours, so it doesn't look like we will ever get those machines assembled. There are about thirty-six people running the operation, and only a couple of men to obey their orders."[1]

Still, there was a war to be won, and progress readying Marine DH-4s inched ahead. By August 17, MacLeish could report, "Our old land hacks are almost ready now. We got one engine in today, and it should be ready to fly by Tuesday." Two days later he predicted, "I think the first D. H. will be ready tomorrow." When he finally lifted off in one of the machines he exulted, "I flew the Liberty [Liberty-powered DH-4] that we were supposed to have and it sure is a wonder. Talk about power. I haven't seen much more. She seemed very solid. She stunted beautifully too."[2] Even with this welcome development, it would be weeks before the first machine reached the Marines and many more weeks after that until enough aircraft were delivered from Eastleigh to permit independent operations to commence.

In fact, Pauillac readied only a few DH-4s for the Northern Bombing Group. Instead, preparations soon shifted to the Group's principal assembly and repair facility at Eastleigh in southern England. Even as aviators served with various RAF squadrons, labored over balky Caproni aircraft, or attempted to expedite work at designated aerodromes, their compatriots in Britain rushed to complete construction of the vital assembly and repair base. With shortages of aircraft looming as the program's Achilles' heel, placing the facility in operation assumed greater importance with every passing day.

The Navy acquired the uncompleted RAF acceptance park at Eastleigh in late July. Despite extensive construction already under way, much work remained to be done to accommodate the two thousand aviation personnel scheduled to occupy the site. Lt. Godfrey Chevalier served as the first commanding officer. He arrived from Dunkirk July 23, accompanied by the public works officer, Lt. Frederick Bolles of the Civil Engineer Corps. A paymaster and medical officer showed up shortly thereafter followed by a few bluejackets. The Navy men lacked almost everything necessary to complete the facility, including barracks, mess halls, building supplies, tools, even typewriters for office paperwork. Chevalier described trying to wring "order and intelligent service out of chaos and inexperience. . . . I am trying to prepare all hands for the avalanche that is coming."[3]

Chevalier and his charges pushed ahead, however, and slowly order appeared out of seeming chaos. Additional staff and personnel poured in. The first labor draft reached Eastleigh in early August. A construction department that grew to seven hundred men raced to lay pipe, string power lines, and build large galleys and mess halls. Machine shops opened. Disassembled aircraft and engines arrived. Marine officers disembarked from France to hurry the process along. On September 21, Cdr. Bayard T. Bulmer assumed permanent command, imposing a more nautical tautness to operations. Machine shops went on double shifts to compensate for shortages of tools and equipment. The first Liberty-powered DH-4 assembled at Eastleigh reached France October 2, ten weeks after the Navy occupied the site. Regular ferrying operations began soon after.

In the second week of October, the base assembled six Liberty motors and three aircraft, two DH-9As acquired from the British and one DH-4 shipped from the United States. Major Cunningham, commander of the Day Wing, assigned Capt. William McIlvain to expedite the process of transporting the machines to France. By the end of the month, Eastleigh was turning out several battle-ready aircraft per week, despite persistent shortages of machine tools and other necessary equipment and a devastating visitation of influenza that claimed nearly twenty lives. At the time of the Armistice, 220 officers and 2,200 enlisted personnel worked at the facility.[4]

With aircraft so slow to reach the field, many Marine aviators served initially with RAF squadrons. The first three pilot-observer crews joined 218 Squadron August 9, 1918, just a few days after reaching the front. This group included pilots Robert Lytle, Arthur Wright, and Herman Peterson, and observers Donald Whiting, Charles Needham, and William McSorley. In the weeks that followed, Lytle/Whiting participated in eight raids and dropped 1,600 pounds of bombs. Peterson/McSorley and Wright/Needham each carried out four attacks, both crews dropping 800 pounds of ordnance. Targets included Ostend, Zeebrugge, Bruges, and Westend. On one early mission, Peterson became separated from his formation and wound up landing 120 miles south of his home aerodrome. According to Lt. Charles Todd, a squadron mate who compiled both a detailed personal diary and a "Brief Outline of the Work Accomplished by Squadron 'C,' " flying these [RAF] missions meant encountering "the prize archie batteries of the German Army and Navy . . . a barrage of shrapnel and projectiles that was so heavy that it seems not humanly possible to pass through it." They returned from their RAF duty on August 20.[5]

The next day, another three crews joined 217 Squadron. After completing three missions, the Americans rotated back to their own aerodromes, replaced by three new crews. Beginning in early September, six leatherneck pilots entered the RAF Pilots Pool at Audembert where they carried out practice flights on DH aircraft.[6] Following qualification, they would be forwarded to front-line squadrons. As originally envisioned, all inexperienced aviators would pass through this facility. Pilots ordered to report on September 5 included Sidney Clark, John McMurran, Peter Lawson, and Marcus Whitehead. A few Marines also flew with 202 Squadron, a reconnaissance unit. Before war's end, 48 pilots, 14 observers/aerial gunners, and 43 enlisted ground crew served with the British, about 10 percent of total FMAF strength.[7]

The first lone Marine aircraft reached Flanders via Pauillac on September 7, 1918.[8] In the following weeks, a few more Liberty powered DH-4s made their way to northern France. These began arriving from Eastleigh October 2. Modifications to make the machines safe to fly occasioned the long delay. A joint American-British technical committee inspected the U.S.-manufactured airplanes and mandated changes that required large quantities of materials not immediately available. As one of the official histories noted, "Much delay was experienced in obtaining this material as it was of a priority type." The amount of effort required to rework the machines also necessitated lengthy post-modification testing. Not until well into October were these various obstacles overcome.[9]

Delays transporting and preparing DH-4s for Marine use caused Captain Cone to commence negotiations with RAF authorities looking to exchange Liberty engines for surplus British-manufactured DH-9 airframes. During 1918, British factories produced 32,000 airframes but only 22,000 engines. Atlee Edwards, Sims' aide for aviation, believed "we might drive a very good bargain with the British to exchange for them airplanes without motors."[10] He believed up to 40 Handley Pages and 290 DH-9s might be available. The discussions eventually resulted in allocation of 54 machines, which were delivered to Eastleigh for assembly and testing.[11] Mated with American-built Liberty engines, the British DH-9 was reclassified DH-9A. They, too, arrived in France slowly and in small numbers, resulting in missions incorporating a mix of aircraft, though with similar capabilities.[12]

The Allied offensive in Flanders that opened on September 28 dramatically increased the tempo of air operations and several Marine pilots attached to 218 Squadron joined the fray.[13] Major Cunningham proudly related that the British had been very complimentary about the Americans' performance. Flying RAF aircraft, Marine crews "made numbers of raids every day and did quite a lot of low flying and trench strafing. They got into a number of very severe fights." On September 26, 2nd Lt. Chapin C. Barr was attacked while on a mission over enemy territory.[14] Although mortally wounded—a bullet had severed an artery in his leg—Barr drove off the enemy. He died shortly after returning to the aerodrome due to the extensive loss of blood. It was believed Barr brought down an enemy machine before being wounded. Two days later on September 28, Lt. Everett Brewer and his observer, GySgt. Harry Wershiner, battled an enemy fighter. The fierce encounter left pilot Brewer with a bullet wound "through the fleshy part of his leg and the fleshy part of his buttock." Observer Wershiner was shot through the arm and lung, but survived. They received official confirmation for destroying one German machine and perhaps a second.[15] In yet another encounter that day, 2nd Lt. Frank Nelms Jr. dueled with a German fighter, as did Capt. Francis Mulcahy, and 2nd Lt. Chapin Barr.[16] The following day, Captain Mulcahy and his observer GySgt. Thomas McCullough brought down a German triplane.[17]

As the Allies' Flanders offensive unfolded, some ground units moved forward more rapidly than others, and by October 1 French and Belgian infantry had become isolated near Stadenburg, cut off from food and supplies. Belgian aviation headquarters immediately organized an emergency food drop, utilizing many aircraft, including those from 82 and 218 Squadrons. Up to ten individual rations were placed in bags filled with dirt and loaded in the airplane cockpits, to be dropped by hand at specified points. Upon impact, the bags burst open, with the rations cushioned by the dirt packing. Over a period of four hours during the morning of October 2, up to 80 aircraft dropped 15,000 rations weighing more than 13 tons.[18]

Three Marine aircraft piloted by Mulcahy, Capt. Robert Lytle, and Lt. Frank Nelms participated in this mission, conducting five food-dropping sorties. Carrying bread and canned meat, they descended to heights of 100 to 300 feet, in many cases braving heavy enemy ground fire, to "deliver" emergency rations. Each pilot received the Distinguished Service Medal for this exploit. Their observers, Sgt. Archie Paschal, Sgt. Amil Wiman, and Sgt. Thomas McCullough were awarded the Navy Cross.[19]

Lacking sufficient bombers to conduct independent missions, Marines sometimes flew their own aircraft as part of raids carried out by 215, 217, and 218 squadrons, a true example of amalgamation. According to Captain Hanrahan, the tender of all aircraft in commission to 5 Group, was "accepted very gladly." A previous attempt to offer the Marine units to the AEF had been politely declined. This practice of joint operations continued until mid-October when the Americans secured enough machines to begin launching their own attacks, though still at targets selected by British planners.[20] In essence, the Marine Day Wing functioned as an element of 5 Group. The first American-built DH-4 with a Liberty engine and American-manufactured guns participated in a raid against the railroad yards at Lichtervelde on October 1, dropping four 112-lb bombs on the target.[21]

Flying with 218 Squadron during the opening days of the Flanders offensive constituted the Marines' true baptism of fire, relentlessly attacking railways, troop trains, and harbor installations. They suffered their first battle casualties and deaths. Cunningham reported after a week of combat, "All the pilots deserve considerable credit for this strenuous and dangerous work. . . . A number of the machines have been shot up rather badly." In all, leatherneck personnel participated in 43 missions while with the British, dropping more than 15,000 pounds of bombs. Maj. Bert S. Wemp, commanding officer of 218 Squadron, praised the Americans for their wonderful work, lamenting, "When the time comes for those now here to leave the Squadron, I can assure you that I will hate to part with their services."[22]

The Marines also came away deeply impressed with their RAF hosts and their redoubtable leader. According to Karl Day, the major "taught me what it means to be an officer and a gentleman. He was a remarkable commanding officer." American flyers were welcomed to the unit, but as newcomers frequently filled the spot on the vulnerable trailing right edge of the V flight formation, the British reasoning that if the Marine got shot, they had not lost anything.[23]

The last of the Marine day units, Squadron D, counting 42 officers and 183 enlisted men under the command of Lt. Russell Presley, reached La Fresne October 5.[24] The combined strength of the Day Bombing Wing then stood at 159 officers and 842 enlisted men. Despite great difficulty acquiring bomb racks and other vital equipment, the number of serviceable aircraft inched upward. These activities coincided with the resumption of heavy fighting along the Flanders front. Marine headquarters noted, "The present offensive in this sector started last Monday [October 14] and we have been making strenuous efforts to help it out from the air and have done very well." This meant Day Bombing Wing units were now launching their own missions in their own aircraft.[25]

On Monday, October 14, No. 9 Squadron conducted its first independent raid, utilizing a mixed force of 10 DH-4 and DH-9A aircraft to hit the railway yards and sidings at Thielt, Belgium, with 2,218 pounds of ordnance. Typical loads included two 112-lb and two 50-lb bombs. One machine— 2nd Lt. Donald Newell Whiting—failed to get off. Another, piloted by Lt. Harvey Norman, returned to base with engine trouble. At least one aircraft landed with a 50-lb bomb precariously attached to the wing as the balky release had failed to work. With Captain Lytle in the lead, they approached the target in a V formation. The Marines carried out the attack in "very poor" visibility from an altitude of 15,000 feet.[26]

On the way home, a swarm of a dozen enemy Fokker and Pfalz fighters jumped the group and firing commenced at distances ranging from two hundred to four hundred yards. Four enemy aircraft focused on Lytle's aircraft, while eight others went after a DH-4 piloted by Lt. Ralph Talbot and crewed by crack shot Corp. Robert Robinson. In the ensuing fight, Robinson downed one attacker, but sustained severe wounds, including a shattered elbow and knee and two or three gut shots. Clearing his jammed guns with one hand, he continued firing until he passed out. Hugely outnumbered, Talbot maneuvered violently and eventually downed another German scout. He then dove toward the ground, passing over the German lines at an altitude of barely fifty feet. He landed at a Belgian aerodrome near Hondschoote, where the unconscious Robinson received emergency care at a field hospital and survived his severe wounds. Both Talbot and Robinson later received the Medal of Honor for their actions.[27]

While Talbot and Robinson battled for their lives, flight leader Lytle attempted to come to their aid but his engine failed, forcing him to make a long, unpowered glide back toward friendly territory. Passing over the German trenches at an altitude of 1,000 feet, he set his machine down 200 yards inside the Belgian lines behind a railway embankment, stopping just short of a large shell crater. Lytle and his observer jumped from their craft uninjured. A group of Belgian soldiers dragged the downed aircraft into the relative safety of the shell-hole and then they and the Marines crawled over the bank and into protective dugouts on the far side. First reports transcribed at headquarters at 2 p.m. simply stated that Lytle and Wiman were missing, with their machine, No. D-3, "last seen near Westrude travelling northwest." Word they were safe reached Autingues at 4:45 p.m.

With so few aircraft available, however, the Marines were not about to write off Lytle's machine. Under the eyes of enemy troops, a work party disassembled the airplane, hauled the pieces across open ground, and loaded them on board a truck and trailer. Enemy artillery fire showering the truck and men with shrapnel hastened their nighttime departure from the battlefield. The men and aircraft returned safely to their aerodrome where the machine was ultimately cannibalized for spare parts.[28]

Before the Armistice brought operations to a halt, Nos. 8 and 9 squadrons carried out fourteen raids, dropping a total of 18,792 pounds of bombs. With the anti-submarine mission eliminated, the Marines became a tactical force, "hindering as much as possible the retreat of the enemy in this sector." Typical targets included railway junctions and yards, canals and canal locks, supply dumps, and aerodromes located at Thielt, Steenbrugge, Eecloo, Ghent, Deynze, and Lokeren. On several occasions, dreadful flying conditions forced aircraft to turn back from their objectives or abort the mission entirely.[29]

Prior to each raid, NBG headquarters prepared a detailed Operations Order that identified the target, named an alternate objective, and provided appropriate map coordinates, as well as specific instructions. Fears of possible damage to Belgian civilians or economic assets ranked high among concerns, as did worries about "friendly fire" incidents. A mission planned against Steenbrugge railway junction for October 17 included the admonition, "Care must be taken to bomb the main junction and not any spur tracks into the works. . . . No bombs are to be dropped within 2,500 metres of our front lines. Bombs will be dropped from as low an altitude as A. A. defenses permit."[30] Similarly, instructions for the planned October 18 raid against the Melle railway junction and sidings stated explicitly, with the text underlined for emphasis, "Under no circumstances are bombs to be dropped west of the line from Bruges to Thielt."[31]

Upon receipt of the Operations Order, Day Wing headquarters issued a specific Raid Order that summarized the operations directive and added further information, including the distance from the aerodrome to the primary and secondary objectives and references to available reconnaissance photographs. Officers were directed to phone the results of the mission to wing headquarters as soon as the last aircraft landed at the field or within one hour of the return of the first machine if one or more planes were unaccounted for.[32]

Squadron commanders continually stressed the importance of maintaining tight formation during a raid and closely adhering to the flight leaders' instructions. Though sometimes faster than their German opponents and more heavily armed, the Marines' DH-4 and DH-9A aircraft were less maneuverable than the German Fokker D.VIIs and Albatross D-Vs they faced. In addition, while the DH-4 could climb higher than its antagonists, the DH-9A could not. Enemy aircraft hunted in packs, often of a dozen or more, but proved extremely reluctant to close with massed day bombers. Should one or two machines be cut off from the group, however, the Germans would pounce on the isolated airplanes.[33]

The mission planned against Melle junction for October 19 revealed the importance accorded flight discipline. Shortly after 10 a.m., six aircraft from Squadron 9 set out on a raid, but turned back when they encountered dense low clouds near the Belgian lines. According to the squadron's commander, Maj. Douglas B. Roben, one Navy pilot, Lt. Albert Johnson, "did not recognize [the] abandon raid signal on this morning's raid. He continued to proceed over enemy territory." While Roben commended Johnson's bravery, "for the sake of maintaining discipline," he wished to ground the impetuous lieutenant for two or three days and replace both him and his observer with another crew "as part of a disciplinary action."[34]

The only combat deaths experienced by the Marines as an independent force occurred October 22 when seven German aircraft attacked Lieutenant Norman's machine over Belgium. The mission, led by Capt. Karl Day, began badly. Nine aircraft set out early in the morning headed toward Ghent, but ran into thick fog as they crossed the lines. The five rearmost planes turned back, leaving just four to carry on, who formed up in a diamond pattern. According to Lt. Charles B. Todd Jr., "The fog was very dense and we had much difficulty in seeing the earth a great part of the time." Intense anti-aircraft fire bounced the machines around like kites in a windstorm. German fighters attacked out of the mist. Three jumped Norman's and Todd's machines in succession, and then flitted away, wary of the fog and Marine guns and the risk of being hit by their own Archie. In the confusion, the small American formation scattered. A large group of German aircraft then picked off Norman. Driven down out of control, both he and observer 2nd Lt. Caleb Taylor died in the ensuing crash near the Bruges-Ghent canal. Todd later reported, "The Hun vultures returned and flying low over the wreckage discharged burst after burst from their machine guns at the two fallen men."[35]

A few days later, Ralph Talbot, a budding poet who had recently escaped an ambush by a dozen enemy aircraft, died in a fiery crash at the beginning

of a test flight at La Fresne. Second Lt. Colgate W. Darden, future governor of
Virginia and president of the University of Virginia, miraculously survived
the crash. Talbot's aircraft, badly shot up in his dramatic aerial encounter
of October 14, had been undergoing repair. On October 25, he pulled the
machine out of the shop for a test flight, inviting his friend Darden along for
the ride. Rather than buckle up, Darden simply sat on the belt. Talbot revved
the engine and they ran down the field, noting the Liberty power plant performed poorly. Talbot taxied back to the starting point and attempted to
take off again. The airplane, now moving at a good clip, lifted a few inches
off the ground before it ran into a bank of earth at the end of the field
thrown up during excavation of a bomb storage pit.

The landing gear of Talbot's DH-4 hit the bank and the aircraft flipped
over and crashed into the pit, catching fire. Talbot died in the flames, or perhaps was crushed by the large fuel tank located behind his seat. Darden, without his seat belt, went flying, "just like a stone being thrown out of a catapult,"
he recalled many years later, and ended up far from the wreck in an open
field. The ground crew, what Darden called the "squadron boys," rushed in
to free the flyers and remove the bombs before the whole place went up. One
chronicler noted, "It was impossible to try and extricate Lieutenant Talbot due
to the terrific heat from the burning plane. First Sergeant John K. McGraw
showed remarkable bravery in the manner in which he took command of the
men, and it was undoubtedly due to his capable leadership in assembling the
men and removing the bombs that the entire camp was not blown up and
many killed." The frantic crew assumed Darden was dead, along with Talbot,
as no one had seen him launched from his seat during the crash. Sometime
later they found him lying in the field, severely injured, with the right side of
his face crushed, a dislocated spine causing temporary paralysis, and a gaping
flesh wound on his left leg, but alive. Darden was rushed by car to a nearby
British hospital in Calais and thence to London for a lengthy convalescence.[36]

The Marines' final combat losses occurred October 27.[37] Late in the
morning, they launched a raid against the railroad junction at Lokeren.
Completing the bomb run about 11:30 a.m., the formation turned to the
north. At 9,000 feet, Lt. Frank Nelms' machine developed engine trouble
and began losing altitude but he attempted to stay with the squadron.[38] Four
enemy aircraft appeared, though keeping a safe distance. Observer John
Gibbs fired a red flare to warn the others. The flight leader followed the
ailing aircraft down until Helms ordered his observer to fire another signal
light indicating they were "washing out."

By now, Nelms' engine was missing badly. Somewhere below 1,500 feet his propeller stopped. With the enemy closing, and machine-gun fire winkling up from the ground, Nelms headed north toward Holland, gliding into a plowed field near Schoodijke, seven or eight miles inside the Dutch border. With the propeller stopped in the vertical position, the airplane nosed over. Shaken but uninjured, the Americans surrendered their weapons and flying gear to local authorities and were interned until the end of the war. They were repatriated November 14.[39] This raid and a second carried out the same day against Ghent, proved to be the Marines' final offensive missions. With the German army rapidly withdrawing from much of Belgium, the Corps relocated part of its aviation force to the recently abandoned enemy aerodrome at Varssenaere and later to Knesselaere near the Dutch border.

Since "Idle hands are the Devil's workshop," Major Cunningham made every effort to keep his men busy and productive despite shortages of serviceable aircraft. Pilots and observers received instruction in map reading, photography, formation flying, bombing, and machine-gun work, sometimes in improvised classrooms or, if weather and equipment permitted, utilizing the Marines' limited supply of aircraft. When not carrying out construction and maintenance duties at their own aerodromes, enlisted personnel were assigned to neighboring RAF squadrons where they participated in the full range of activities at those operational units.[40]

Continuing to emphasize the dominant role played by the British in shaping American practice and organization, Hanrahan claimed, "The training and operations of the Intelligence, Mapping, Photographic, and Aerographic Departments have been developed along lines suggested by the experience of the Allied air forces." Staff officers heading each of these departments had been "afforded every facility by the Allied forces in this sector for obtaining information, and for observing operations of their respective departments."[41]

Marine aviation units, like virtually every other military formation on the Western Front, suffered heavily in the great Spanish Influenza pandemic that struck in 1918. Major Cunningham reported October 31, "The influenza epidemic has played havoc with us for the last two weeks . . . today I tried . . . to form a raid and found that I could only muster ten well pilots out of all four squadrons. . . . We have not enough well officers to fly and not enough well men to handle the machines. I have practically had to abandon operations." Day Squadron 9, the heart of the leatherneck force, first felt the disease's hot breath October 25 when 30 men fell ill. The next day, the number of cases exploded to 102. The YMCA hut became a hospital for officers, with the aerodrome's most severe cases transferred to British facilities in Calais. Three officers and five enlisted men died, including Maj. Douglas Roben.[42]

The death of Roben, a veteran of Vera Cruz and Santo Domingo, hit particularly hard. "This command," noted Cunningham, "without exception, are cut up about the death of Major Roben." He had fallen ill but refused to go to bed until it was too late. Cunningham called Roben, "the best squadron commander I had and I consider that he sacrificed himself to his work." And then, as quickly as it had come, the epidemic passed. By November 9, Cunningham could report, "I am glad to say that the influenza epidemic . . . seems to be leaving us. We have very few new cases and most of the old ones have recovered sufficiently to go to work."

The end was now in sight. In early November, Cunningham wrote, "Today the Armistice negotiations were started." Rather than responding with joy, he lamented, "I am very much afraid that before we can get any more work, the hostilities will cease. Since we have just received enough machines to operate on a good scale, I certainly hope that the Germans refuse the armistice, for at least a few weeks, until we can show just what we can do. . . . However, I am afraid it will be our luck for the Germans to accept the armistice and knock us out of the opportunity." The guns fell silent on November 11.[43]

Chapter Eighteen

Bombing, Bombing, and More Bombing

E ven as the Northern Bombing Group struggled to get off the ground and into action in the summer and fall of 1918, Navy planners seemed determined to continue and expand such activities. In fact, as early as the winter of 1917–18 the subject of long-range bombing was "in the air" and the Navy had caught the scent. Over the next year, it developed or contemplated four major initiatives that, had they been completed, would have deployed many hundreds of aircraft and thousands of personnel, including: creating a significant, land-based bombing force in Flanders to conduct continuous day and night raids against U-boat facilities in Belgium—the Northern Bombing Group; establishing a seaplane base in northern England from which to launch long-range attacks against principal German fleet anchorages—NAS Killingholme; organizing a massive land bomber and seaplane force in Italy to attack Austro-Hungarian military assets at Pola, Trieste, Fiume, Cattaro, and elsewhere—the Southern Bombing Group; and constructing a vast fleet of high-speed "sea sleds" capable of carrying multiengine aircraft within range of their targets and then launching them. All were designed to take the fight directly to the hornets' nest.

During the summer of 1918, staff members in the office of the Director of Naval Aviation undertook a careful analysis of the progress being made within the NBG program preparatory to a major conference with the British scheduled for mid-September to discuss the future of the Navy's bombing plans. They reviewed the origins of the program and its experiences to date, and then made recommendations for upcoming action. The authors defined the mission as operating against submarine bases using land machines, going on to note, "This program is strongly favored by aviation forces abroad, as concurred in by the British and French and the War Department, all agreeing that it is a naval undertaking, except possibly the War Department, which has partially agreed."

On the issue of whether the sputtering program should be continued, eliminated, or expanded, great weight was accorded the fact that "nearly all of the material for the establishment of the stations has been shipped overseas, and flying fields and a repair base have been secured. The personnel is there, ready to go ahead. . . . The only equipment lacking is machines in sufficient numbers." The question of accommodating Army objections also remained on planners' minds. "Ever since the original program was modified," the analyst noted, "and land machines were substituted for seaplanes for these squadrons, the Department has been concerned lest these operations should encroach upon Army prerogatives." Several messages had flowed from Washington to Admiral Sims cautioning that "operations must be confined to purely naval activities."

In fact, the Department always approved of the objective, but worried about using land machines to accomplish it and whether there were sufficient numbers available. The issue hinged on whether bombers employed in naval work qualified as naval aircraft. This, in turn, would determine allocation of scarce Liberty engines that had become integral to completion of the entire initiative. In the Department's view, using both night and day bombers to destroy submarine bases, thereby preventing U-boats from operating, qualified as "purely naval work." Thus, day and night land machines operating against submarine bases were employed in a naval mission and therefore entitled to be classed as naval aircraft.

No matter what the jurisdictional debate or current operational shortcomings, however, a basic question remained: "Are we going ahead with it?" Here the answer was unambiguous, revealing much about the Navy's commitment to newly embraced bombing tactics. Not only would the present program be continued, it should be expanded significantly. Department analysis declared, "We have put our hands to the plow, and we cannot turn back. A tremendous amount of preparation has been done, and our intentions have been made known to the world. . . . This project may lead towards a united air service, but the argument for amalgamation would be ten times as strong if we fail to accomplish our mission."

Not surprisingly, the final recommendations called for MORE! The size of the Northern Bombing Group should be increased to 6 squadrons each of day and night bombers, fielding a force of 168 aircraft. All possible Liberty engines drawn from Navy inventories and elsewhere should be shipped overseas. A field for land plane training should be secured in the United States, separate from the Army Air Service system. Finally, "Be prepared, in so far as possible for further expansion of bombing operations to accomplish the mission of preventing enemy submarines from operating." The Department had spoken, giving bombing a nearly unqualified endorsement.[1]

Testifying before the General Board in late summer 1918, Cdr. John Towers even advocated utilizing massive NC flying boats then under development, to bomb Berlin as well as German naval bases. This, however, was a decidedly unrealistic notion. Flying boats were extremely vulnerable to anti-aircraft fire. Where, exactly, would a stricken machine set down should mechanical trouble develop while crossing over land. A broad river or very large lake would not always be available. No matter, only a lack of suitable aircraft and the unexpectedly swift end of the war prevented some of these plans from being implemented.

While Washington reviewed the NBG situation with an eye toward the future, a similar process unfolded in Europe. The strong overseas support referenced in the Department's analysis flowed from several sources and in many cases extended beyond raids on naval vessels to encompass attacks against factories, dockyards, and even civilians. In August, Lt. Harry Guggenheim at Paris headquarters advised Captain Cone that the Navy's future aviation program should mandate bombing submarine bases first, with convoying operations and coastal patrols ranking second and third in importance. His well-thought-out recommendations encompassed strategic attacks on a wide range of targets, including ports, dockyards, and naval craft, as well as factories producing naval material and "enemy organizations for defense of any of these."[2]

Lt. Cdr. W. Atlee Edwards, Admiral Sims' aide for aviation and an early Pensacola flyer who never received his wings, pushed for a 200 percent increase in NBG strength. He recommended "that the General Allied Air Policy for 1919 be made more offensive in character," adding, "There is no doubt that land aeroplanes are more effective for bombing enemy Naval Bases from the East Coast of Italy and from the North of France than are seaplanes." He also touted the virtues of steam-driven turbines to power aircraft and the qualities of the immense Handley Page four-engine bomber then undergoing flight testing.[3]

Lt. Allan Ames, serving on Sims' staff, also weighed in. A charter member of the First Yale Unit, he labored as a flight instructor at NAS Bay Shore and passed through the RFC gunnery school in Texas before receiving training in two-engine flying boats at NAS Hampton Roads. In July 1918, he joined Paris headquarters, assigned intelligence and operations duties. His work included frequent visits to air stations in France. Ordered to London August 24 he performed similar inspection duties throughout England, Ireland, and France.[4] Writing home in September, he advised his friend Trubee Davison, "Get behind the bombing idea as much as you can; big boat work has been very much overrated from what I have seen here. . . . Big boats are all right for some work, but our resources in naval aviation could be applied in an infinitely more effective way."[5]

Some of the strongest calls for increased strategic bombing emanated from Lt. Cdr. John Callan in Rome. Intimately familiar with the ongoing Italian bombing campaign utilizing heavy Caproni aircraft, he had long advocated American attacks against Austrian naval targets at Pola. He now endorsed a huge program designed to cripple enemy activities in the Adriatic. Significantly, Callan specifically included civilian attitudes as a legitimate target, noting, "The above program would break down the morale of the civil population, driving them from the towns, thus weakening manufacturing strength and fostering revolt." These sentiments concerning civilians as targets were in line with ideas advanced by Lovett and McDonnell in March and April and were integral to the expanding bombardment program carried out by General Trenchard's Independent Force.[6]

Similar support emanated directly from the top. Based on a lengthy review conducted during the summer, Admiral Sims on August 25 recommended suspending seaplane operations at NAS Dunkirk in favor of American-manufactured Liberty DH-4s flying from NBG fields. The Department approved his request September 12.[7] During the same period, preparations continued for a series of naval policy conferences to be held with the Allies regarding plans for operations in 1919. These included soliciting the ideas of naval aviation's principal commanders. According to one historian, "The consensus, as could be foreseen, was in the line of extending our bombing work greatly, both in the German and Austrian areas. This was to be done with land-based planes." A detailed agenda prepared for the upcoming Anglo-American meeting endorsed continuous expansion of bombing activities even at the expense of increased patrolling.[8]

Recommendations generated at the September conference included increasing the size of the Northern Bombing Group by 50 percent, adding four new squadrons. If this program achieved its objectives of destroying the enemy submarine complex in Belgium, the units would then attack "naval objectives if practical, or any other objective suitable to the furtherance of the aims of the military forces." A long-range bombing campaign directed against German forces and installations in the Heligoland Bight also received support. Sims' September 25 report of the conference to CNO Benson endorsed strengthened bombing efforts in Flanders, the Adriatic, and the Heligoland Bight, missions to be undertaken by long-range, land-based bombers.[9]

From his new post in London as Admiral Sims' aide for aviation, Captain Cone also beat the bombing drum loudly. In a "Dear Irwin [Noble]" letter written after the conclusion of the latest conference, he also called for abandoning seaplane patrolling and increasing the size of the Northern Bombing Group. He claimed that "the unanimous opinion with reference to our future [1919] effort is that any additional undertakings should lie along the lines of bombing enemy naval bases, first in the Adriatic and later as soon as suitable machines can be obtained, those in the Heligoland Bight." He accepted as axiomatic that all future bombing "will necessarily have to be done with land machines," specifically those based in England that could carry out the mission and then return to their home aerodromes. He was likely referring to the new Super Handley Page V/1500 long-range aircraft just then being deployed for the first time.[10]

The Department ultimately embraced many of Sims' and Cone's recommendations, agreeing that naval aviation's principal efforts should be focused on bombing enemy naval targets, particularly submarine bases. Plans to attack German bases near the Heligoland Bight would commence, along with related necessary aircraft production. Washington remained skittish, however, about antagonizing the Army and directed that any bombing conducted in the Adriatic region be carried out by seaplanes only.

Even as discussions regarding future actions took place, engineers, mechanics, and aviators worked feverishly to solve technical deficiencies that thus far had crippled the heavy bomber program, especially attempted modifications to flawed Caproni aircraft and power plants. Mechanics at St. Inglevert labored to improve the safety and reliability of Fiat engines, but with little success, and could never complete the required four-hour running test. Various reports documented poor workmanship and construction of the original engines. Lacking adequate facilities in the field, Lieutenant McDonnell, Lt. (jg) Reginald Coombe, and Machinist Mate William Miller, all of them veterans of flights across the Alps, ferried one of the Capronis to Eastleigh. Crews there disassembled the engines, re-machined some parts, and then put them all back together. One of the now properly fitted and assembled power plants then ran for eight hours. No matter, none of the original Capronis ever re-entered service. A plan to equip later bomber deliveries with far more reliable Isotta Fraschini motors would have solved almost all of these issues, but the first example did not reach Eastleigh until November 8.[11]

In their search for an effective large bomber, both the Army and Navy also attempted to correct the Capronis' structural defects. During the summer of 1918, the Bureau of Aircraft Production's special mission based in Paris investigated the aircraft's deficiencies and made recommendations to improve safety and reliability. Led by banker Henry Lockhart Jr., the members met with officials in France and Italy, toured the Caproni factory, and interviewed Army and Navy personnel associated with the program, including Maj. Fiorello La Guardia and several Navy pilots. Their findings documented the many problems already encountered in the field. They also offered numerous suggestions to be instituted in both Italian and American production facilities for the large number of aircraft expected to be manufactured in coming months.

Discussions with Navy and Army pilots at Milan, for example, described the propensity of Fiat engines to backfire and catch fire, often while flying. One American pilot stated that "the Italian officer [Gabriele] d'Annunnzio would not allow any of his pilots to cross the lines in machines equipped with Fiat motors." Others claimed that many crashes occurred because of the actions of "heavy pilots," men with relatively little training in these large machines. Mechanical defects included weak landing gear and inadequate shock absorbers, weak bracing for the rudders, and gasoline feed types subject to frequent breakage. The investigators also uncovered a very "rough and inaccurate system of sand testing" the machines at the factory. More lengthy and specific information documenting the Capronis' shortcomings were gleaned from information supplied by the French Section Technique Aeronautique.

Recommended changes included substituting Isotta Fraschini for Fiat engines in aircraft procured in Italy. A laundry list of structural improvements included stronger wheels and reinforced axles, improved bracing of wings and rudders, stronger wing spars, reinforced wing struts and fuselage, a modified gasoline feed distribution system, and stronger cross wiring of wings. More rigorous safety testing and greater safety factors should be required of all machines. No Liberty engines should be installed until the aircraft had been "thoroughly redesigned and its strength proved satisfactory by sand loading."

The reason that such extensive reworking of an inadequate machine was even considered can be found in an extract from the report of the Lockhart Mission dated October 9, 1918, only one month before the Armistice. Simply put, "There is no night bomber developed sufficiently superior to the 'Handley Page' and 'Caproni' to warrant any change in present production program." In other words, the United States was committed to a massive bomber production program. Many hundreds of machines were expected to operate in 1919. Nothing better was available. Let production proceed.[12]

Fearing technical and production problems plaguing the Caproni program might never be solved, the Navy kept a watchful eye on development of the Super Handley Page V/1500. As early as June 3, 1918, Aide for Aviation Edwards wangled "a short hop" on one of these giant aircraft. He called it "a wonderful machine," with 40 already under production. Mr. Handley Page claimed his newest bomber could carry 6,000 pounds of ordnance, 10 machine guns, and a crew of 7 for up to 15 hours, reaching a speed of 100 mph and an altitude of 17,000 feet. At the end of July, Edwards announced that flight trials would commence in a few days and he would try to arrange a visit by Assistant Secretary of the Navy Roosevelt. He enthused, "It certainly is a revelation to one who has not seen these gigantic machines."[13]

Edwards continued to monitor the progress of the program. In an August missive to Cone he observed that the concept of bombing German naval bases "is a sound one, but must be done from land machines and will, I think, eventually be successful when machines of the Super Handley Page type have been adopted." One example was then being fitted with four 500-hp engines. Anticipating that the Navy might acquire some of these giants, he predicted, "With such a machine as this I think we could look forward to the time when we will be able to bomb German Submarine Bases successfully."[14]

The envisioned land-based aerial assault against targets in Flanders constituted just one aspect of the Navy's many-faceted embrace of a bombing offensive. What became known as the Killingholme-lighter program grew out of discussions held in London back in October 1917 between Cdr. Hutch Cone and Admiralty officials. The prospect of attacking the German fleet directly in its lair through the use of destroyer-towed lighters proved tempting. Such a lighter had already been suggested at Felixstowe by Cdr. John C. Porte, a concept that evolved into a rather substantial vessel, fifty-eight feet in length and displacing twenty-four tons, sufficient to carry flying boats weighing up to five tons. The lighters would incorporate an airtight trimming tank in the stern that could be flooded by means of high-capacity pumps and emptied with compressed air. By lowering the stern, flying boats could be launched or retrieved through use of a trolley/cradle and winch.

According to Ens. Joseph Eaton, "The lighters were built so that they could let water in the back of them, and this would make the tail end of the lighter sink and the bow of the lighter point upwards, and they set the plane on a little carriage on a track, and they could let it back down the track and slide into the water, then they'd blow the water out with compressed air and they'd go on their way." Aircraft would then take off from the water. The first sea trials occurred in June–September 1917. The Admiralty planned to build fifty such craft and asked the U.S. Navy to operate a complementary base supporting thirty additional lighters and forty planes (both American built) on the east coast of England, with preparations to be completed by March 15, 1918. Early discussions considered establishing the U.S. effort at Lowestoft, but attention soon shifted to Killingholme instead.[15]

Following discussions in London, Cone and other officers inspected RNAS Killingholme located on the Humber estuary to assess its suitability. Nothing definite resulted at first, but when CNO Benson visited Europe in November he approved the general proposal and cabled Washington to begin construction of necessary flying boats at once. Such plans represented a decisive shift away from defensive patrol activity along the west coast of France to more active, offensive missions based in England and Flanders. The Navy Department selected Lt. Cdr. Kenneth Whiting to head the American portion of the project. He received assurances the initiative would receive top priority.[16]

After some debate, the Navy chose the Curtiss H-16 flying boat for the bombing campaign. Make no mistake, by the standards of the day these were huge aircraft, far larger than the Gotha G.Vs that had terrified London, measuring 46 feet in length, with a 96-foot wingspan. Two 360-hp Liberty motors provided power, generating a top speed of 95 mph. With an empty weight of 7,400 pounds and loaded weight of 10,900, the H-16 carried a crew of 4, radio equipment, 4 or 5 machine guns, and up to 500 pounds of bombs. The plane could remain aloft for 4 hours at maximum power or up to 9 hours at cruising speed. Production would be split between the Curtiss factory in Buffalo and the newly constructed Naval Aircraft Factory in Philadelphia.

Official determination of a suitable aircraft, however, was not made until January. The most likely alternative, the British Felixstowe F-5 flying boat, was even larger, fully as big and heavy as the Handley Page bomber. Ultimately, the Navy decided to ship completed Curtiss products to Europe "as is" and carry out any necessary modifications overseas. Naval Aircraft Factory versions incorporated requisite changes before being shipped. Work on the lighters moved more quickly. Government shipbuilding crews at the New York Navy Yard manufactured the vessels.

Gathering and training necessary personnel required careful coordination. A nucleus of pilots already existed in Europe, with many of them now dispatched to British bases for further training and to gain actual war experience. Others soon arrived from the United States. Back home intensive instruction commenced for additional pilots, radio operators, and engineers at a variety of military and industrial facilities. Forty crews attended the RFC gunnery school established at Fort Worth, Texas, in the winter of 1917–18. Other engineers and mechanics trained at the Packard factory in Detroit (Liberty motors), Savage Arms Works (machine guns), Delco (radios), and the Philadelphia Navy Yard. Still others traveled to New York to learn about the lighters. From Texas and elsewhere pilots moved on to Hampton Roads to study large flying boats.

A small draft of enlisted personnel sailed for England in late February, quickly reinforced by 300 more. After numerous delays, Whiting, along with eight officers and 150 men, and a mountain of aircraft, supplies, and miscellaneous gear reached Killingholme May 30. The Navy complement eventually rose to 94 officers and 1,324 bluejackets, making Killingholme the largest operational station in Europe. The base extended across 135 acres of low-lying, riverside ground. Eight hangars had been built, including the gigantic No. 6, a 10-bay structure measuring 800 feet in length and 220 feet in depth. Two 800-foot slipways constructed of stout planks, iron girders, and piles driven into the riverbed carried aircraft from the concrete apron, across the wide Humber mudflats, and down to the water.

Working the bugs out of the H-16s took time, however. Ens. George F. Lawrence, for example, claimed radiators "practically fell to pieces" on flights of more than thirty miles. Other defects included faulty oiling systems and soft cranking gears. According to Lawrence, aircraft arriving from the United States proved totally unfit for war service and had to be practically rebuilt inside, including shifting fuel tanks, constructing machine-gun mounts, installing bomb gear, and adjusting controls. Nearly every machine exhibited a curious twist that made it tail heavy and wing heavy, imparting a tendency to turn left. Crews worked on this for months but never fully succeeded in getting it fixed. Well into September only a half dozen planes were operable. New radiators did not arrive until October 1.[17]

And what of the great lighter project, original justification for the entire effort? Toward the end of February 1918 Royal Navy forces from Harwich mounted an exercise in the North Sea. Destroyers towed lighters into position. Rough water and high winds prompted personnel to spread oil on the water to calm the waves. The first airplane slipped off the lighter easily enough but oil soon covered the windscreen, pilots' goggles, and compass. Several more exercises followed and work continued throughout the spring and summer but with little to show for it, and in July the British decided to terminate the initiative.

The following month, Admiral Sims notified Captain Cone of the change in plans. "Operations with towing lighters," he noted, "have not proved very satisfactory." Weather conditions could not be forecast accurately, resulting in three or four failures for every successful action. Lighters, destroyers, and support ships faced risks not justifiable if the objective could be attained otherwise. Additionally, anti-aircraft defenses had been improved considerably, and "Large America" flying boats were vulnerable to attack.

By contrast, long-distance, land-based bombers capable of self-defense had been developed for either day or night action, while Sims liked the concept of aircraft carriers much better than seaplanes. "Accordingly," he reported, "the Admiralty do not now contemplate long distance bombing operations in any large scale with seaplanes from towed lighters and have diverted some of their towed lighters from Felixstowe to northern bases to act as temporary accommodations for "Large Americas," which had proven useful for long-range reconnaissance and attacks on enemy mine sweepers and light forces. The American commander suggested Killingholme's lighters be retained for that purpose and personnel trained accordingly.[18]

A few days later, Edwards added, "I believe conditions have so altered in the North Sea since the time the Killingholme project was conceived that there is absolutely no use attempting to operate lighters in that area." Bombing German bases was a good idea, but required land planes, especially Super Handley Page 1500s then under development. He concluded, "I cannot see any possible chance of our utilizing seaplanes or flying boats for this purpose." Perhaps the lighters should be used around the coast of Ireland instead. Even after the British abandoned the program, sporadic experiments at Killingholme continued.[19]

But as the weeks passed filled with unresolved H-16 problems any possibility of attacking Kiel in 1918 grew more and more remote. As late as October, however, some proposed putting the original plan into action, but bad weather and rough seas made experiments impossible. Efforts to secure scarce American destroyers for these experiments also failed. Nonetheless, Killingholme personnel kept trying and might have attempted a raid had not the Armistice intervened.

Back in the United States, Cdr. Henry C. Mustin tackled the same challenge, how to conduct long-range bombing raids against German naval bases with aircraft that could both reach the target and then make their way home safely. During the early weeks of the war, Secretary Daniels had broadcast a service-wide radiogram calling for new and innovative ideas. Mustin responded on August 19 with a bold plan, recommending a sustained aerial assault against German fleet and submarine facilities, both military and civilian. He listed four objectives: destruction through continuous bombing of the submarine shops at Emden and Wilhelmshaven; torpedo attacks on submarines and other vessels in their home bases; torpedo attacks on German fleet elements at sea; and destruction of the Essen works.

Mustin acknowledged current aircraft possessed neither the range nor lift capacity to complete such missions. Instead, he proposed constructing thousands of small, high-speed seaplane carriers—based on a design by Canadian Albert Hickman—that could quickly and independently transport bombers and their escorts within striking range, each capable of launching one fully armed aircraft. He described the unusual vessels as a "type of seaplane which has, in effect, a double set of floats; one set is for cruising and planing and remains in the water when the aeroplane takes the air; the other set, permanently attached to the airplane, is for landing and flotation only," in essence a large hydroplane speedboat/aircraft hybrid. The combination of aircraft engine power, boat speed, and wind would achieve flying speed. By transporting the aircraft close to the enemy shore and then using high speed and natural wind to effect the launch, "The bombing trip to Germany and return to England could thus be made with the gasoline carried by the plane."[20]

Despite deficiencies in the plan that he frankly acknowledged, Mustin called for 1,200 "sea sleds" to carry fighting or scouting machines, 2,000 more to transport single-engine bombers, and 2,400 sleds to carry large bombers or torpedo planes with a 2,000-lb load. The huge numbers were meant to facilitate continuous raids and accommodate expected attrition. The strike forces would assemble in England and motor across the North Sea before launching their attacks. Alternatively, Texel Island, off the Netherlands coast, might be utilized as a base, launching swarms of bombers from the waters of the Zuider Zee. A later biographer called the proposal "wildly ambitious." Nonetheless, the visionary aviator assumed the nation's substantial automobile industry could carry out construction.[21]

The fall and early winter of 1917–18 witnessed a flowering of the Navy's interest in projects promising a means to strike directly at the enemy. The Department approved the Killingholme project and both the General Board and London Planning Section of Admiral Sims' London headquarters soon examined the issue as well. On December 31 Cdr. John Towers, the number two man in naval aviation and real heart of the program, declared Mustin's proposal "seems to be feasible but more definite computations should be furnished. . . . In order to accomplish this I recommend that Commander Mustin be ordered to temporary duty in Operations, Aviation Section, and the Bureaus of Construction and Repair, and Steam Engineering be requested to complete and check the plans, cooperating with Commander Mustin in this work. This model can be undertaken from present general plans."[22]

Mustin's visionary plan, similar in some ways to the towed-lighter program being developed for Killingholme, offered one very realistic advantage. The Killingholme initiative envisioned use of relatively slow H-16 or F2A flying boats, which carried comparatively small bomb loads and possessed a limited service ceiling. By contrast, plans for Mustin's sea sleds eventually included employing Caproni bombers, with considerably greater speed, higher service ceiling, and the ability to carry substantial bomb loads, more than 1,600 pounds under ideal conditions. These aircraft would be able to deliver far larger quantities of ordnance, faster, from safer altitudes

Mustin reported to the Bureau of Construction and Repair on February 26, 1918, after service as executive officer on board the battleship *North Dakota*, his new assignment to oversee an experimental sea sled development program. With the support of the chief of the bureau Rear Adm. David Taylor, Mustin pressed ahead. On May 27, the Navy contracted with the firm of Murray & Tregurtha, Inc. to design and manufacture the sea sleds along lines developed in cooperation with Taylor and Mustin.[23]

Difficulties obtaining Caproni bombers with which to conduct tests slowed the process. In mid-July, Admiral Taylor notified the Bureau of Aircraft

Production of the need for two Caproni biplanes by October 1. Efforts to acquire machines from the Standard Aircraft Company had thus far proven unsuccessful. Taylor therefore asked if the Navy could have two of the small number of aircraft Standard had promised to the Army.[24] At an "informal conference" held a few days later, representatives of the Bureau of Aircraft Production promised to supply one machine and attempt to secure another from the Italian Mission.[25]

Only one day later, however, the assistant director of the Bureau of Aircraft Production formally notified Taylor that two American-made Capronis could not be supplied. Throwing his support behind the project to "use Caproni airplanes on board detachable motor lighters," Secretary Daniels then contacted Secretary of War Newton Baker directly. He explained, "In view of the fact that if this experiment proves successful, a very successful weapon will be developed for attacking submarine bases, and since the department has used every endeavor to obtain the necessary two lighters on which these airplanes will be carried, and since these lighters will be ready in the course of a short while, it is requested that immediate delivery of the two planes be effected." Not until the fall, however, could the Navy secure an airplane for experimental purposes.[26]

Work on the project continued throughout 1918, centered at the experimental station at Hampton Roads. By early autumn, Murray & Tregurtha had delivered the first two vessels, each 55 feet long and driven by four 450-hp engines. Naval constructor Jerome Hunsaker worked directly with Mustin to modify a Caproni bomber for use on board the novel craft, which could reach speeds of 55 mph on their own and 60 mph with the bomber's engines running at full throttle. The project was terminated shortly after the Armistice when an attempt to launch a large bomber from one of the sleds failed.

As Cdr. Pat Bellinger, the commanding officer at NAS Hampton Roads, described the event, "It was close quarters on the sea sled, and everybody had to have a working job, and had to be in his place. All engines were started, plane and boat, and we worked up to full speed heading into the wind. The idea was to reach flying speed of the Caproni, then release it from the sled. Great disappointment. We never quite reached flying speed for the Caproni, so we had to call the test a failure." Mustin later testified that problems with the aircraft's release mechanism prevented a successful launch. The following March, however, he succeeded in getting a much smaller Curtiss N-9 training aircraft airborne. According to Bellinger, "To my way of thinking, this was the Navy's first airplane carrier."[27]

A proposal to create a large Southern Bombing Group based in Italy to attack critical Austro-Hungarian military facilities along the Adriatic coast constituted yet another attempt to institute a strategic bombing program. A small-scale effort was already in the works, at the new naval air station at Porto Corsini, which carried out several raids against the Austrian fleet base at Pola.[28] In mid-August 1918, Lt. Cdr. John Callan, commander of Navy aviation efforts in Italy, submitted a lengthy report endorsing an enormous enlargement of America's Mediterranean efforts. Offering an overtly strategic perspective, he believed decisive results could be achieved on the Italian front.

Presaging Winston Churchill's "soft underbelly" concept of World War II, Callan predicted additional pressure would precipitate collapse of the Hapsburg Empire, eliminating Germany's strongest ally, isolating Bulgaria and Turkey, thus fracturing the Central Powers. To achieve a knockout blow, he urged continuous bombardment of enemy installations at Pola, Trieste, Fiume, and elsewhere. Callan specifically recommended establishing a large seaplane base at Vallona equipped with H-16 flying boats to mount attacks every night, while simultaneously operating a huge squadron of eighty Capronis at Poggio Renatico to conduct continuous bombing of the Austrian fleet base at Pola. Callan also urged the Navy to assume control of the existing air station at Ancona.[29]

Callan endorsed attacks against a wide range of targets, including submarine bases, coast defenses, air stations, arsenals and repair bases, fleets, munitions factories, and merchant ships in harbors. His proposals recommended committing great numbers of personnel and supplies, and perhaps North Sea–type towed lighters to permit long-range reconnaissance missions, possibly supplanted by fast cruisers converted into seaplane ships. Callan's program envisioned the stupendous total of 932 aircraft, dwarfing the Northern Bombing Group project. Not surprisingly his proposals sparked vigorous debate in London and Paris and extensive negotiations with the Italian government. On September 19, Edwards at Navy headquarters cautioned the huge project was not deemed wise and that existing commitments should be carried out first.

Nonetheless, both Sims and Cone endorsed the program at such time that sufficient aircraft became available. That the idea was considered more than pie in the sky became evident when Cone notified Captain Irwin on September 24, "I am now collaborating with [Captain Luke] McNamee and [Captain Frank] Schofield [both of the Planning Section][30] with a view of getting up a cable home with reference to our future effort in the Adriatic."

Cone also urged Irwin to come to Europe "in order that you may be on the ground to close any arrangements we may make and thus back us up when it comes to the matter of fulfilling our engagements, and not be cut down on our undertaking as was done in the case of the Northern Bombing Project." The events of late May and early June still rankled.[31]

Despite Edwards' rebuff, when a commission headed by Cdr. Benjamin Briscoe visited Italy in early October to investigate aircraft supply issues, Callan won them over to his plan. Following submission of their findings, Admiral Sims' headquarters authorized exploratory discussions in Rome. Edwards, however, remained wary, believing Austria-Hungary tottered near collapse, while the bombing project would take many months to implement. Nonetheless, preliminary activities began. A small Navy group headed to Venice October 8 and inspected the island of Poveglia in the lagoon. The Navy also considered occupying the nearby islands of Sacco Sesola and Malamocco. Additional preparations included meetings between Naval Attaché Charles Train and the Italian Ministry of Marine. Eventually word arrived that a new commission of five officers would visit in early November to examine the plan in detail. On October 23, Edwards informed Cone (then recuperating from severe injuries sustained in a torpedo attack), "I have organized quite a little mission to proceed to Italy and report to Capt. Train for the purpose of investigating the feasibility of our going into the Adriatic area." The group included Kenneth Whiting as head of the commission, along with A. W. K. Billings, Charles Mason, H. H. Lane, Harry Guggenheim, and Spenser Grey. "They will leave tomorrow night and will, I feel sure," Edwards noted, "establish a certain amount of prestige for us both by their numbers and by their rank."[32]

Anticipating approval, officials in Italy commenced arrangements, ordering a hangar moved into the region and preparing bunks for several hundred men. Captain Train recommended transferring ninety flying boats and forty land planes to various Italian facilities. He also urged dispatch of a civil engineer, an Assembly and Repair officer, and a senior Bureau of Supply officer, along with sufficient bluejackets to begin work. Train's actions coincided with dispatch of the inspection commission, but the sudden collapse of Austria-Hungary caused them to cut their journey short in Paris. The Southern Bombing Group remained just a glimmer in Commander Callan's eye.

All these plans and proposals—Killingholme, Southern Bombing Group, Mustin sea sleds, expanded Northern Bombing Group operations— shared several characteristics. Each embodied the aggressive, offensive spirit that defined presidential and naval aspirations, if not activity, during the war. All embraced radically new technology and tactics, well beyond the Navy's experience or charter. Each required an enormous commitment of resources—financial, material, and personnel—yet each received strong support from the highest echelons of the Department. The Navy entered the war with a newly constructed battle fleet of dreadnoughts but no way to employ them. It ended the conflict embracing a strategic bombing mission intended to harass the enemy across an aerial front that stretched from the Adriatic to Flanders to the North Sea.

Chapter Nineteen
Lessons and Legacies

Actual war service of the Northern Bombing Group consisted of several related, but distinct, elements. Plagued by insurmountable shortages of aircraft, the Night Bombing Wing spent most of the period from August to November 1918 on the sidelines, constructing aerodromes and attempting frantically to modify their inadequate Caproni machines. The unit's most concrete contribution came through provision of flight and ground personnel to a British Handley Page squadron where they serviced aircraft and participated in numerous raids. Another small band of aviators attached to the Northern Bombing Group flew with British and French squadrons, carrying out a variety of missions during the final push in Flanders, ranging from bombing raids to ground attacks and offensive patrols.

The failure of night bombing squadrons to conduct active wartime missions caused little but frustration and heartache. These units were expected to provide the Northern Bombing Group's principal hammer to smash the German submarine complex in Belgium. They alone represented strategic airpower in its purest form. But all the training and aerodrome construction could not overcome a simple fact: the group lacked adequate numbers of bombers, and those they obtained proved fatally flawed, incapable of combat operations. The Northern Bombing Group's Day Wing, manned almost entirely by Marine aviators, also experienced frustration and delays caused by slow delivery of warplanes. Many spent time with RAF squadrons, learning their trade and supplying much-needed reinforcements to hard-pressed British units. Not until October 14 did the Americans possess sufficient aircraft to mount missions of their own. By this time, the Germans were evacuating their submarine facilities and the Marines turned instead to attacking railroad yards and sidings, attempting to disrupt the enemy withdrawal.[1]

Unfortunately, despite so much effort, progress proved more apparent than real. At the time of the Armistice, the four Marine squadrons possessed just fifteen serviceable aircraft. Most of their combat service came while attached to British units. Eastleigh did not become a productive facility until October. Aerodrome construction continued throughout the period, with much of it still uncompleted when the guns fell silent. And in a final irony, the very mission that underlay creation of the NBG evaporated when dramatic Allied advances in early fall caused Germany to abandon its submarine bases in Belgium. Only the work of a small group of Marine aviators and the exploits of David Ingalls, the Navy's first "Ace," brightened the overall picture.

Naval aviation entered World War I without a mission or doctrine, though with a commitment to marrying the airplane to the fleet. During the war, it pursued many divergent paths, largely determined by the desire to get the submarine: a massive investment in flying boat technology and procurement; huge numbers of coastal patrol stations in the United States, Europe, and elsewhere; and plans for a gigantic offensive program starting with the Northern Bombing Group. Unfortunately, the Navy was unable to fully implement its strategic plan before the war came to an unexpectedly abrupt end in November 1918.

Nevertheless, the Northern Bombing Group constituted an important aspect of aviation development in the U.S. Navy. Though historians of Marine Corps aviation date the origins of that activity to Alfred Cunningham's beginning flight training in May 1911 it was only World War I and leatherneck service with the Northern Bombing Group that gave substance to the early visions and predictions. The exploits of the Corps' four day bombing squadrons provided the foundation upon which all future growth and accomplishment rested.

The Northern Bombing Group program also demonstrated the size and significance that a naval aviation program could assume. Training efforts stretched across two continents. Huge facilities arose in France and England. Hundreds of aircraft and many thousands of personnel were gathered and organized. The work of nearly all the bureaus and offices of the Navy Department played a role in the development and execution of the plans, as did the highest echelons of the service's leadership. Had the related programs involving the Killingholme lighters, Mustin sea sleds, and Southern Bombing Group been carried out, the Navy would have amassed an offensive weapon of startling size and power. In the end, however, the bombing initiative proved to be "the road not taken." Despite all the work and planning, the Navy abandoned its efforts in land-based aircraft after the Armistice in order to concentrate on taking aircraft to sea on board its ships. Thus, the NBG represented an important part of the crucial process by which the Navy figured out what to do, and what not to do, with the airplane.

The Navy's success in adopting the airplane in World War I was due in large part to the foresight of its leadership, starting with Secretary Daniels and CNO Benson, debunking the often-proclaimed myth of the "overly conservative" battleship admirals. It also tells us much about technological innovation in the Navy and its application to warfare during this era. Admittedly, the infatuation with strategic bombing yielded few tangible results. Nonetheless, the bomber interlude offers several points to ponder.

It underscored the open-endedness of naval aviation's mission and promise in the days before the aircraft carrier ruled the fleet. The bombing initiative highlighted the Navy's eagerness to get into the fight by any means, to experiment, to embrace a new, largely untried weapon, and to commit enormous technical, material, and personnel resources to its development and use. Planning and implementation of the varied programs emphasized the interdependent intertwining of American and British aviation programs, both intellectually and physically. The turn to land-based bombers also underscored the general unsuitability of large seaplanes for major combat operations, and thus provided a strong nudge toward the aircraft carrier as the solution for taking airpower to sea. Finally, the sharp competition for material between the Army and Navy as exemplified by the Caproni controversy contributed to the nasty postwar aviation rivalry between the services.

World War I witnessed the birth of strategic bombardment, a doctrine and strategy that eventually assumed enormous importance in both the Royal Air Force and the United States Army Air Forces. Inter-war claims regarding aerial bombardment's impact on civilian morale surfaced initially in 1917–18, however. And, ironically, the first American military force to move aggressively to implement the new doctrine was the U.S. Navy. This has not been fully acknowledged before. Though strategic bombing had little impact on the course of World War I, it was where air warfare was going, and the sailors were there first.

The Navy's disappointing experience attempting to field a credible bombing force, and the virtually insurmountable difficulties encountered, spotlighted such efforts' utter reliance on the fields of technology, engineering, production, and logistics. As events clearly revealed, without a strong, innovative scientific, engineering, and manufacturing base, no program was possible. It is no coincidence that Robert Lovett, the young reserve officer who developed much of the Northern Bombing Group program in 1918, only to be stymied as commander of the Night Bombardment Wing by the failure to obtain serviceable bombers, brought that lesson with him when he returned to Washington in 1940. As Assistant Secretary of War for Air, he worked feverishly to organize, modernize, and expand the American aviation manufacturing sector and partnered with Gen. Henry "Hap" Arnold to develop the strategic bombing campaigns unleashed against Germany and Japan.

The Northern Bombing Group of 1918 had been, in large part, Lovett's brainchild and he never abandoned the idea. In a 1942 interview, he described his World War I experiences, saying, "I got tired of chasing submarines all over hell. . . . The way to get 'em was to pound their bases from the air until reduced. . . . The successful use of this weapon depends on its employment en masse, continuously and aggressively. . . . That's what we've got to do now." He added that "our main Job is to carry the war to the country of the people who are fighting us."[2] Both literally and figuratively, it was not very far from the World War I raids on the submarine bases at Bruges and Zeebrugge to the World War II skies above Berlin and Tokyo.

Notes

Introduction
1. Neville Jones, *The Origins of Strategic Bombing*, 19.
2. Robin Higham and Stephen Harris, eds., *Why Air Forces Fail*, 354.
3. Edward Coffman, *The War to End All Wars*, 197.

Chapter 1. Blazing the Path
1. Morrow, *The Great War in the Air*, 43.
2. Jones, *The Origins of Strategic Bombing*, 48.
3. Alexander, *The War in the Air*, vol. 1, 265–6.
4. According to Pollard, tracer ammunition was not introduced until late 1915 when Buckingham tracer was bought by the RNAS (see "The Royal Naval Air Service in Antwerp, September–October 1914," 3).
5. Hedin, *The Zeppelin Reader*, 88.
6. Lea, *Reggie*, 22–4.
7. Marix claims they proceeded to take the Tabloids home with them, but the official record cards, according to Pollard (see note 4), show that the two Tabloids did not arrive at Eastchurch until September 9, 1914.
8. Pollard, "Royal Naval Air Service in Antwerp," 13.
9. Lea, *Reggie*, 26–27.
10. Gilbert, *Churchill: A Life*, 283–4.
11. Lea, *Reggie*, 27; Bruce, "The Sopwith Tabloid, Schneider and Baby," 736; Pollard, "Royal Naval Air Service in Antwerp," 12.
12. Report from Cdr. Spenser D. A. Grey to the Director of the Air Department, Admiralty, on the Raid on Cologne and Düsseldorf, October 17, 1914, reproduced in Roskill, *Documents Relating to the Naval Air Service*, vol. 1: 1908–1918, 179–80. Note: The dates in Grey's report are apparently in error as German sources and official history give the date of the bombing as October 8, and not the 9th.
13. Report from Cdr. Spenser D. A. Grey to the Director of the Air Department, Admiralty, on the Raid on Cologne and Düsseldorf. Grey was later told that

214 | Notes to Pages 11–16

the airship sheds were on the other side of the Rhine River, east of where he was searching.

14. Pollard, "Royal Naval Air Service in Antwerp," 14.
15. H. A. Jones, *The War in the Air*, vol. 2, 350.
16. Whitehouse, *The Zeppelin Fighters*, 78; Bacon, *The Dover Patrol*, 224.
17. Jones, *The Origins of Strategic Bombing*, 59; Ash, *Sir Frederick Sykes and the Air Revolution*, 41, 66.
18. Morrow, *The Great War in the Air*, 113; Jones, *The Origins of Strategic Bombing*, 60; Ash, *Sir Frederick Sykes and the Air Revolution*, 66.
19. HQ RFC Memo. July 1915, AIR 1/921,204/5/889, NAUK, as cited by Jones, *The Origins of Strategic Bombing*, 61–2.
20. Jones, *The Origins of Strategic Bombing*, 66.
21. Charles P. Bartlett, *In the Teeth of the Wind: The Story of a Naval Pilot on the Western Front*, 8.
22. Bruce, "The Sopwith 1½ Strutter," *Flight*, September 28, 1956, 545.
23. H. A. Jones, *The War in the Air*, vol. 2, 430–1.
24. Ibid., 442; Neville Jones, *The Origins of Strategic Bombing*, 82.
25. Jones, *The Origins of Strategic Bombing*, 82.
26. H. A. Jones, *The War in the Air*, vol. 2, 442–3.
27. Reynolds, *Admiral John H. Towers*, 102.
28. Ibid., 103.
29. Ibid., 104.
30. Marine Bernard Smith, assistant naval attaché in Paris at the time, also went over the lines with the French.
31. Nos. 4 and 5 Wings grouped their aircraft into two squadrons according to role (bombing or fighting). Each squadron was subdivided into two six-plane flights, the fundamental fighting unit of the RNAS at the time. The maximum number of bombers that could be put in the air at one time was only twenty-four. Note: The bomb load of the G.IV (depending on the source) is listed at 100 or 113 kg (220 or 249 lb).
32. Thetford, *British Naval Aircraft since 1912*, 286.
33. Bruce, "Handley Page 0/100 and 0/400," *Flight*, February 27, 1953, 259.
34. Bartlett, *In the Teeth of the Wind*, 32; RN Service Record.
35. Robertson, "No. 207 (Bomber) Squadron," *Flight*, October 12, 1933, 1023.
36. 207 Squadron RAF Association, "The Squadron in the First World War, 1917 HPO 0/100s," www.207squadron.rafinfo.org.uk.
37. Jones, *The Origins of Strategic Bombing*, 77–8.
38. No. 3 Wing Royal Naval Air Service was originally formed in the Aegean from 3 Squadron RNAS on June 21, 1915, and was disbanded on January 18, 1916.
39. Dodds, "Britain's First Strategic Bombing Force," 1; Neville Jones, *The Origins of Strategic Bombing*, 123.

40. H. A. Jones, *The War in the Air*, vol. 2, 452.

41. Thetford, *British Naval Aircraft since 1912*, 301; "Sopwith 1½ Strutter Single Seat Bomber," http://www.roden.eu/HTML/404.htm.

42. The French contribution consisted of four Farmans, one Sopwith, and one Nieuport; the British sent two Sopwith bombers and one Sopwith fighter.

43. Dodds, "Britain's First Strategic Bombing Force," 1–2; H. A. Jones, *The War in the Air*, vol. 2, 453; Neville Jones, *The Origins of Strategic Bombing*, 81, 111, 113. The attack on Oberndorf was a joint raid conducted with the French involving thirty-four French and twenty British aircraft.

44. Jones, *The Origins of Strategic Bombing*, 106.

45. Cooper, "The British Experience of Strategic Bombing," 52.

46. Ibid., 85–6. For more details of the dispute see pages 86–90.

47. Ibid., 52.

48. Jones, *The Origins of Strategic Bombing*, 107.

49. Minutes of a meeting with Colonel Barrés [*sic*], held at the Admiralty, October 22, 1916, as cited by Jones, *The Origins of Strategic Bombing*, 91–2.

50. Admiralty to Air Board, October 26, 1916, as cited by Jones, *The Origins of Strategic Bombing*, 92.

51. Henderson to Air Board, October 31, 1916, as cited by Jones, *The Origins of Strategic Bombing*, 92.

52. Jones, *The Origins of Strategic Bombing*, 93.

53. Haig to War Office, November 1, 1916, as cited by Jones, *The Origins of Strategic Bombing*, 94.

Chapter 2. Crushing the Hornets' Nest

1. See Gilbert, *The First World War*, 308, 328. Crippling losses continued into May and June, with 285 and 286 ships sunk, totaling 1,264,000 tons.

2. Still, *The Crisis at Sea*, 12; Freidel, *Franklin D. Roosevelt*, 289. See also Klachko, "William Shepherd Benson: Naval General Staff American Style," 312.

3. For an insider's view of activities in Washington and the Navy Department during this period see E. David Cronon, *The Cabinet Diaries of Josephus Daniels, 1913–1921*, February 1–April 2, 1917, passim; Daniels, *Our Navy at War*, 19–29; and Daniels, *The Wilson Era*, especially 15–39. The full impact of the German submarine campaign remained a closely guarded secret in Britain, with public announcements designed to bolster morale rather than provide accurate information. See also, Trask, *Captains and Cabinets*, 35, 43, 45–6, 49.

4. Daniels, *Years of War and After*, 28–9.

5. Cronon, *Daniels' Diaries*, 116.
6. On March 22, the French attaché reported a similar American initiative. Assistant Secretary Roosevelt's activities in the weeks leading up to the outbreak of war are covered in Freidel, *The Apprenticeship*, 287–300. See also Daniels, *Our Navy at War*, 30–5, and *Years of War and After*, 39.
7. For discussions of the slide toward war see Cronon, *Daniels' Diaries*, 117–27; Still, *The Crisis at Sea*, 1; Coffman, *The War to End All Wars*, 92–3; and Trask, *Captains and Cabinets*, 52–6. See also Lord, "The History of United States Naval Aviation, 1898–1939," 285, 290–290a.
8. Daniels, *Our Navy at War*, 39.
9. Dispatch of a high-ranking officer resulted at least in part from a cable authored by Ambassador Walter Page in London. Adm. Charles Badger, president of the General Board, strongly endorsed Sims for the mission. See Sims, *The Victory at Sea*, 3, 4. See also Cronon, *Daniels' Diaries*, 121–2; Daniels, *Years of War and After*, 65–8; Freidel, *The Apprenticeship*, 300.
10. Cronon, *Daniels' Diaries*, 134.
11. Quoted in Sims, *The Victory at Sea*, 43, 373. See also, Still, *Crisis at Sea*, 13–15.
12. Cronon, *Daniels' Diaries*, 123; Coffman, *The War to End All Wars*, 104–5; Still, *Crisis at Sea*, 25.
13. Cronon, *Daniels' Diaries*, 139; Daniels, *Our Navy at War*, 40–3.
14. Sims, *The Victory at Sea*, 24–5.
15. Trask, *Captains and Cabinets*, 61, 65–6.
16. Cronon, *Daniels' Diaries*, 133, 136. See also Coffman, *The War to End All Wars*, 108–9, and Still, *Crisis at Sea*, 16–17. See especially, Joseph Taussig, *The Queenstown Patrol, 1917*, passim.
17. Trask, *Captains and Cabinets*, 85, 87; Still, *Crisis at Sea*, 17; Daniels, *Our Navy at War*, 130–1, 144–5.
18. Still, *Crisis at Sea*, 18; Trask, *Captains and Cabinets*, 91, 93–4, 95, 131.
19. Trask, *Captains and Cabinets*, 147; Still, *Crisis at Sea*, 19 and 523–4n100.
20. A full description of this event and lengthy excerpts from Wilson's speech are contained in Daniels, *The Wilson Era*, 42–5, and *Our Navy at War*, 145–8. See also Coffman, *The War to End All Wars*, 97; Klachko, *Admiral William Shepherd Benson, First Chief of Naval Operations*, 80; and Trask, *Captains and Cabinets*, 132.
21. James Bradford, "Henry T. Mayo: Last of the Independent Naval Diplomats," 269. Wilson maintained this negative view of the British for many more months. See Klachko, *Benson*, 81, and Trask, *Captains and Cabinets*, 133. On Wilson's view of the Royal Navy's timidity see Daniels, *Our Navy at War*, 143–5.
22. Trask, *Captains and Cabinets*, 137–8, 147. Secretary Daniels later reported that Wilson had employed the hornets' nest metaphor even before the United States entered the war. See *Our Navy at War*, 144.

23. Trask, *Captains and Cabinets*, 147; Bradford, "Henry T. Mayo," 269.
24. Klachko, *Benson*, 81. Mayo's trip to Europe is covered thoroughly in Still, *Crisis at Sea*, 72–4, and Bradford, "Henry T. Mayo," 269–70.
25. Klachko, *Benson*, 82–3; Trask, *Captains and Cabinets*, 149, 150, 156. See also Bradford, "Henry T. Mayo," 270.
26. Klachko, *Benson*, 90.

Chapter 3. Naval Aviation Enters the Arena
1. Quoted in Northern Bombing Group Summary, u/d [September 1918], box 909, ZGN, RG 45, NA. See also, Coffman, *The War to End All Wars*, 105. Benson quoted in *Hearings: Naval Investigation*, 66th Congress, Second Session, Washington, D.C., 1921, 1849–51. According to Still, "Neither the secretary of the navy nor the CNO anticipated the rapid deployment of U.S. naval forces to European waters in great numbers." *Crisis at Sea*, 25.
2. Memo to Secretary of the Navy, re: Naval Aeronautics, June 25, 1917, box 1, World War I Collection, Aviation History Unit, NHHC. Towers, Naval Aviator No. 3, assumed control of the aviation desk in the office of the CNO in the fall of 1916 and held that position until superseded by Capt. Noble Irwin in May 1917. Towers remained as Irwin's principal deputy throughout the war.
3. "Appropriation for Aeronautics," *Air Service Journal*, July 12, 1917, 13.
4. Early aviation efforts are summarized in Arthur, *CONTACT!*, 525–9, and W. H. Sitz, *A History of Naval Aviation*, 9–14.
5. According to Whiting, at Benson's behest, after receiving a request from the French Ministry of Marine.
6. Several sources document the activities of the First Aeronautic Detachment, including "Draft History of U.S. Naval Aviation in France: June 5, 1917 to November 1, 1918," and Whiting to Sims, "History of the First Aeronautic Detachment," November 29, 1918, both box 910, ZGU, RG 45, NA. See also Lord, Naval Aviation History [draft], 171. Whiting's recommendation to bomb enemy submarine bases in Belgium followed on the heels of his conversations with British aviation officers in the Dunkirk vicinity.
7. For a brief outline of Whiting's career see Arthur, *CONTACT!*, 12–13, and *Guide to the Kenneth Whiting Papers, 1914–1943*, at Special Collections and Archives, Nimitz Library, USNA. See especially George van Deurs, *Wings for the Fleet*, 18. Van Deurs knew Whiting and interviewed many of his early friends and associates.
8. Regarding Whiting's drinking, Van Deurs noted that "It was later said of him that he was 'brilliant, capable, and fearless, but a bottle made him unpredictable.'" Van Deurs, *Wings for the Fleet*, 120.
9. Ibid., 118–19.

10. Ibid., 119–20.
11. Ibid., 120.
12. Lord, "History of United States Naval Aviation, 1898–1939," 275a–275b. Historian Clifford Lord also knew many of Whiting's contemporaries and included their observations in his study of early naval aviation.
13. Lord, "History of United States Naval Aviation 1898–1939," 348.
14. Arthur, *CONTACT!*, 5. See also, Rossano, *Stalking the U-boat*, 7, 14, 17, 19, 20, 26.
15. Materials relating to Sayles' activities in box 132, GA-1, RG 45, NA. Also Rossano, *Stalking the U-boat*, 12–13, 14, 20.
16. Letter, Conger to McGowan, June 18, 1917, box 2, Papers as Assistant Secretary of the Navy, 1913–1920, FDRPLM.
17. Jackson to BuNav, July 20, 1917, SS/Ga-144, SHM; Jackson to SecNav and Sims to Benson, August 4, 1917, both in Lord, "History of Naval Aviation, 1898–1939," 351–2. See also *Florence* [Alabama] *Times*, October 25, 1974; *Washington Post*, October 4, 1971; and Rossano, *Stalking the U-boat*, 21.
18. Gorrell Testimony, Hearings Before Subcommittee No. 1 (Aviation) by the Select Committee on War Expenditures in the War Department, 66th Cong., 1st sess., 2937–8; Cooke, *Billy Mitchell*, 56–7; "News from the Classes," MIT *Technology Review*, vol. 22, 527.
19. Maurer, *The U.S. Air Service in World War I*, I, 52–4, and Hudson, *Hostile Skies*, 13–15. The activities and recommendations of the Bolling Commission are documented in Gorrell Histories, Section A, vol. 1, "Report of the Organization of the Air Service."
20. "Process Verbal de la Conference tenue au Ministre de la Marine 8 Juillet 1917," SS/Ga-144, SHM.
21. Rossano, *Stalking the U-boat*, 27.
22. Cuxhaven had been the target of an RNAS raid in 1914. It was likely on this trip that he met Commander Spenser Grey for the first time.
23. Report, Whiting to Sims, August 26, 1918, box 461, PA, RG 45, NA.
24. For Mustin's August 19, 1917, proposal see Morton, *Mustin—A Naval Family of the Twentieth Century*, 112.
25. Cone's description of these events is contained in Memo, Cone to Sims, October 6, 1917, box 7, Papers as Assistant Secretary of the Navy, 1913–20, FDRPLM. See also Lord, "History of Naval Aviation, 1898–1939," 388–90. Also Lord, History of United States Naval Aviation [draft], 102–3.
26. Rossano, *Stalking the U-boat*, 168–9.
27. Lord, "History of Naval Aviation, 1898–1939," 391–2.
28. For Benson's trip to France, see Klachko, "William Shepherd Benson," 318. See also Headquarters Log, USNAFFS, November 23, 1917, box 131, GA-1, RG 45, NA.

29. Headquarters Log, USNAFFS, box 131, GA-1, RG 45, NA. On this tour Benson also visited Brest to confer with Adm. Henry Wilson, commander of naval forces in France.

30. Benson quoted in Daniels, *War Years and After*, 88.

31. Klachko, "William Shepherd Benson," 111.

32. See Benson to Daniels, December 3, 1917, and Cone to Pratt, November 24, 1917, both quoted in Lord, "History of United States Naval Aviation, 1898–1939," 392–3. In his Naval Aviation History (draft), 103, Lord reported, "This idea [lighters] was taken up in a large way by Admiral Benson, our Chief of Naval Operations, when he visited England in November."

33. Extract, Letter, Cone to Sims, January 29, 1918, box 52, Sims Papers, LC.

34. Klachko, *Benson*, 112–13; Letters, Taylor to Sims, December 22, 1917, and Sims to Bristol, January 4, 1918, both box 52, Sims Papers, LC.

Chapter 4. The Dunkirk Dilemma

1. Whiting later wrote, "The control of the sea along the Belgian coast for a distance of twenty miles from land rests with the enemy. . . . The sole remaining offensive arm is the bombing and gun seaplane." Whiting to Sims, August 26, 1917. Paymaster Conger informed Adm. Samuel McGowan, Chief of the Bureau of Supplies and Accounts, "We must act without delay. There is actually not a single smile left in the town." Conger to McGowan, June 18, 1917, box 2, Papers as Assistant Secretary of the Navy, 1913–1920, FDRPLM.

2. Whiting's visits to the Dunkirk region and the events that unfolded there are described in a variety of sources, including an undated, handwritten report on Hôtel Meurice stationery; Memo, Whiting to Sims, "History of the First Aeronautic Detachment," November 29, 1917; Whiting, "First Aeronautic Detachment," December 2, 1918, all box 910, ZGU, RG 45, NA. See also "Draft History of U.S. Naval Aviation in France, June 5, 1917 to November 1, 1918," box 910, ZGU, RG 45, NA.

3. Accepted August 8, 1917. Lord, "History of United States Naval Aviation, 1898–1939," 349–50.

4. The Aircraft Manufacturing Company, Ltd. (Airco) DH-4 was designed by Geoffrey DeHavilland and became a mainstay of the British and American aviation services. A later modification, also built by Airco and powered by a 240-hp Siddeley Puma engine, was redesignated the DH-9.

5. See Westervelt to Sims, "Comments regarding United States Aviation Forces," August 17, 1917, and Whiting to Jackson, "Progress in Aviation," u/d [August], both box 461, PA, RG 45, NA.

6. The recently introduced Hansa-Brandenburg W-12 caused much of this carnage. According to John Morrow, its "speed and maneuverability would

give the British a rude shock over the Flanders coast in 1917," in "Defeat of the German and Austro-Hungarian Air Forces in the Great War, 1909–1918," 117, in Higham and Harris, eds., *Why Air Forces Fail.*

7. Whiting to Jackson, August 10, 1917, box 461, PA, RG 45 NA.
8. For the history of early activities at NAS Dunkirk see "Draft History of U.S. Naval Aviation in France," box 910, ZGU, RG 45, NA; Letter, Johnson to Van der Veer, October 12, 1917, box 460, PA, RG 45, NA; Letter, Whiting to Cone, December 10, 1917; Letter, Chevalier to Cone, December 15, 1917; and Weekly Reports, November 17, 1917, and January 19 and 22, 1918, all box 461, PA, RG 45, NA. A description of conditions at Dunkirk is contained in Paine, *First Yale Unit,* II, 141. See also Alonzo Hildreth, "Over There," 56–63.
9. Van Deurs, *Wings for the Fleet,* 77.
10. Arthur, *CONTACT!,* 7. For a description of Chevalier's personality, see van Deurs, *Wings for the Fleet,* 77, 80.
11. Tours, Moutchic, Lac Hortin, St. Raphael, and Issoudun were located in France. Gosport was in England, while Turnberry and Ayr were in Scotland.
12. The slow gathering of flight personnel at Dunkirk is documented in Sheely, *Sailor of the Air;* Rossano, *The Price of Honor;* Moseley, *Extracts from the Letters of George Clark Moseley;* and Rossano, *Hero of the Angry Sky.* Gates' time at Dunkirk is discussed in Paine, *First Yale Unit,* II, 140–9, and several Gates letters written to Trubee Davison in the winter of 1918 in Davison Papers, YA. Training centers at Cranwell, Eastchurch, and Leysdown were all located in England.
13. Paine, *First Yale Unit,* II, 144.
14. German Report, October 20, 1917; Weekly Report, NAS Dunkirk, November 17, 1917, both box 461, PA, RG 45, NA.
15. Whiting to Cone, December 10, 1917, box 461, PA, RG 45, NA.
16. Rossano, *Stalking the U-boat,* 74.
17. Sheely, *Sailor of the Air,* 79, 87.
18. Alonzo Hildreth, "Over There—World War I."
19. The station log records only five missions that month. See Station Log, NAS Dunkirk, RG 24, NA.
20. Paine, *First Yale Unit,* II, 147.
21. Ibid., 195; Moseley, *War Letters,* 157.
22. Moseley, *War Letters,* 157–9.
23. Rossano, *Price of Honor,* 69–70. See Progress Report, July 15, 1918, box 133, GA-1, RG 45, NA, and "U.S. Naval Aviation Fatalities Abroad . . . Dating from April 6, 1917," box 69, Entry 17, RG 72, NA.
24. U.S. Navy, Office of Naval Records and Library, *The American Naval Planning Section, London,* 91–101.
25. Ibid., 101–16.

26. Ibid., 107–16. See also Rossano, *Stalking the U-boat*, 383, nn35–37.

27. Lord, Naval Aviation History (draft), 171.

Chapter 5. Bombardment Aviation

1. Morrow, *The Great War in the Air*, 138, 205.

2. Quoted by Morrow, in ibid., 205.

3. Note: German artillery had been bombarding the city of Reims for three and a half years.

4. Martel, *French Strategic and Tactical Bombardment Forces of World War I*, 142–3.

5. See Holley, *Ideas and Weapons*, 48, and Sweetser, *The American Air Service*, 66.

6. General Commander-in-Chief to Minister of War, May 6, 1917, Exhibit B (Translation) attached to Mitchell, Memorandum for the Chief of Staff, U.S. Expeditionary Forces, June 13, 1917, Microfilm Series M990, Roll No. 8, NA-Md.

7. Mitchell, Memorandum for the Chief of Staff, U.S. Expeditionary Forces, June 13, 1917.

8. White, *Mason Patrick and the Fight for Air Service Independence*, 11; Levine, *Flying Crusader*, 67; Testimony of William Mitchell, U.S. Congress, *Hearing Before Subcommittee No. 1 (Aviation)*, 2616.

9. Hurley, *Billy Mitchell: Crusader for Air Power*, 23; Miller, *Billy Mitchell: Stormy Petrel*, 5.

10. Mitchell, *Memoirs of World War I*, 59.

11. Ibid., 69.

12. Morrow, *The Great War in the Air*, 200.

13. Maurer, *U.S. Air Service*, vol. II, 107; see also Mitchell's memorandum of June 13 referenced above.

14. Gorrell Testimony, 2948.

15. Atkinson, *Italian Influence on the Origins of American Strategic Bombing*, 143.

16. Gorrell Testimony, 2949–50.

17. *Aerial Age Weekly*, November 18, 1918, 512.

18. Bolling to Coffin, October 15, 1917, as cited by Holley, *Ideas and Weapons*, 57.

19. Cooke, *Billy Mitchell*, 57; Miller, *Stormy Petrel*, 9.

20. Tucker, ed., *The European Powers in the First World War*, 766.

21. *International Military Digest Annual 1917*, 14.

22. Tucker, "Gotha G IV Bomber."

23. Jones, *Origins of Strategic Bombing*, 131–2.

24. H. A. Jones, *War in the Air*, vol. 5, 29–31.

25. Williams, *Biplanes and Bombsights*, 38.

26. Ibid., 36.
27. Ibid., 39.
28. Second Report of the Prime Minister's Committee on Air Organization and Home Defense Against Air Raids, August 17, 1917, Appendix II, H. A. Jones, *The War in the Air: Appendices*, 10.
29. Jones, *Origins of Strategic Bombing*, 136–7.
30. Ibid., 135.
31. Williams, "Shank of the Drill," 391.
32. Hardinge Goulborn Giffard Tiverton, Second Earl of Halsbury.
33. Jones, *Origins of Strategic Bombing*, 112.
34. Williams, "Shank of the Drill," 391.
35. Cooper, "The British Experience of Strategic Bombing," 55.
36. Lord Tiverton, "Original Paper on Objectives," September 3, 1917, Halsbury Papers, AC 73/2, RAF Museum, Hendon, UK, as cited by Williams, "Shank of the Drill," 392.
37. H. A. Jones, *War in the Air*, vol. 5, 90; Williams, *Biplanes and Bombsights*, 43.
38. The RFC units were No. 55 Squadron, a day bombing unit flying the DH-4 and No. 100 Squadron equipped with obsolete FE-3bs.
39. RAF "Bomber Command: First Bombers," http://www.raf.mod.uk/bombercommand/bc_devel1.html.
40. Williams, "Shank of the Drill," 395.
41. Ibid., 397.
42. Gorrell to Bolling, October 15, 1917, as cited by Holley, *Ideas and Weapons*, 135.
43. Holley, *Ideas and Weapons*, 135.
44. CAS to AEF Chief of Staff, December 1, 1917, 9, Roll No. 1, Microform Series M990, RG-120, NA-CP.
45. See Williams, "Shank of the Drill," 399, for details.
46. Ibid.; Holley, *Ideas and Weapons*, 24.
47. Edgar S. Gorrell, "History of the Strategical Section," February 1, 1919, 1, Roll No. 10, Microform Series M990, RG-120, NA-CP.
48. "Death of Lt. Col. Spenser Grey," *Flight*, October 14, 1937, 386. According to Grey's official service record, he was hospitalized on August 29, 1917, for injuries from an undisclosed cause (the handwriting on the original is indecipherable).
49. Spenser Douglas Adair Grey, Royal Navy Service Record ADM/196/152, PRO.
50. CAS to AEF Chief of Staff, December 1, 1917.
51. Williams, "Shank of the Drill," 408.
52. Burtt Report to AEF Chief of Staff, January 1, 1918, as cited by Williams, "Shank of the Drill," 408.

53. Gorrell, Memorandum for Chief of the Service, January 2, 1918, reproduced in Gorrell, "An American Proposal for Strategic Bombing in World War I," *Air Power Historian* 5, no. 2 (April 1958), 115.

54. Williams, "Shank of the Drill," 410.

55. Memorandum No. 12, February 15, 1918, reproduced in *The American Naval Planning Section London*, 91–6.

Chapter 6. The General Board Speaks

1. Daniels, *Our Navy at War*, 216.

2. Marder, *From the Dreadnought to Scapa Flow: The Royal Navy in the Fisher Era, 1904–1919*, vol. 5: *Victory and Aftermath (January 1918–June 1919)*, 116–17.

3. A useful summary of Cunningham's aviation career is found in Greg Malandrino, "Alfred Austell Cunningham: Father of Marine Corps Aviation." See also Johnson, *Marine Corps Aviation: The Early Years 1912–1940*, and Arthur, *CONTACT!*

4. For individual biographies see Arthur, *CONTACT!*

5. The Advance Base Force was intended to seize, hold, and defend advanced bases for the fleet, permitting warships to refuel and conduct repairs in time of war while far from permanent facilities in the United States and elsewhere. For a brief summary of the relationship between early Marine aviators and the Advance Base Force concept see Michael Morris, "Combat Effectiveness: United States Marine Corps Aviation in the First World War," 232–3. See also Johnson, *Marine Corps Aviation*, 1–2, 10, and Mersky, *U.S. Marine Corps Aviation, 1912 to the Present*, 2.

6. The highest-ranking officer in the Marine Corps in this period was designated the Major General Commandant. That term is no longer used. It is now the Commandant of the Marine Corps (CMC).

7. Cunningham's tour of Europe is recounted in his detailed diary. See Cosmas, *Marine Flyer in France: The Diary of Captain Alfred A. Cunningham, November 1917–January 1918*. Sailing with him was Adm. Albert Niblack and 50 million dollars of gold bullion.

8. *Conyngham* (DD 58) and *Jacob Jones* (DD 61) were both launched in 1915 and operated out of Queenstown, Ireland, part of the large American commitment of destroyers to anti-submarine patrol in European waters.

9. Cunningham Diary, November 26, 1917.

10. Quoted in Johnson, *Marine Corps Aviation*, 15.

11. To the Commandant and the General Board, January 24, 1918, in box 3, Cunningham Collection, MCL.

12. In addition to Cunningham's written report and his General Board testimony, see his "Memorandum Prepared at the Request of the General

Board Giving Status of Submarine Offensive Recommended by that Board," April 5, 1918, box 2, World War I Aviation Papers, MCL.

13. Whiting testimony, January 16, 1917, in Proceedings of the General Board of the U.S. Navy, 1900–1950, M1493, NA.

14. Albion, *Makers of Naval Policy, 1798–1947*, 78.

15. See Morton, *Mustin—A Naval Family of the Twentieth Century*, 85, 89–90.

16. Cunningham testimony of February 5, 1918, in Proceedings of the General Board of the U.S. Navy, 1900–1950, M1493, NA.

17. Memo, Cunningham to Barnett, February 6, 1918, box 3, Cunningham Papers, MCL.

18. Quoted in Proposed Letter, Second Section, General Board to Secretary of the Navy, February 25, 1918, box 910, ZGU, RG 45, NA. See also box 3, Entry 35 [Records of USNAFFS, Belgian Bombing to France, Misc.], RG 72, NA, and Summary of NBG Correspondence, June 24, 1918, box 910, ZGU, RG 45, NA.

19. Such as the twin-motor H-16, then under development/construction at Curtiss and The Naval Aircraft Factory.

20. Proposed Letter, Second Section, General Board, February 25, 1918, box 3, Records of USNAFFS, Belgian Bombing to France, Misc. (Entry 35), RG 72, NA; see also General Board Report 820 to Secretary of the Navy, Feb. 26, box 6, Entry 23, RG 72, NA.

21. Opnav dispatch 3569 discussed in NBG Summary, box 6, Entry 23, RG72, NA. See also box 909, ZGN, RG 45, NA.

22. Quoted in draft Northern Bombing Group history (not paginated), box 910, ZGU, RG 45.

23. Cunningham received orders to create the escort squadrons March 11. NBG Summary, June 24, 1918; also draft Northern Bombing Group history, both box 910, ZGU, RG 45, NA.

Chapter 7. Paris Charts a Different Course

1. Lovett remains one of the most important figures of the twentieth century who lacks a comprehensive biography. See Isaacson and Thomas, *The Wise Men: Six Friends and the World They Made,* 60–3, 90–3; Paine, *First Yale Unit*, vols. I and II, passim; Wortman, *Millionaires' Unit*, passim; Arthur, *CONTACT!,* 40–1.

2. Letter Report, Lovett to Cone, January 2, 1918, box 155, GU, RG 45, NA. See also Paine, *First Yale Unit*, vol. II, 168–9. Additional analysis of operations at RNAS Felixstowe is contained in Letter Report, Vorys to Cone, February 2, 1918, box 1, Entry 35 [Belgian Bombing to France, Misc.], RG 72, NA.

3. Lovett remarks, Yale Unit reunion transcript (1924), 139, Davison Papers, YA.

4. Lovett's only association with land planes had come in the summer of 1917 when he attended the French primary flight school at Tours.
5. Lovett's meeting with Grey is recounted in Paine, *First Yale Unit*, vol. II, 171–2.
6. Letter, Westervelt to Sims, August 18, 1917, quoted in Lord, "History of United States Naval Aviation, 1898–1939," 372.
7. Grey introduced Lovett to Captain Charles Lambe, the Admiralty's leading proponent of concentrated bombing of submarine infrastructure. Lambe let Lovett review his unit's documents and advised in formulation of a plan. See Wortman, *Millionaires' Unit*, 204.
8. Paine, *First Yale Unit*, vol. II, 171.
9. Griffin letter #1975, quoted in Northern Bombing Group Summary, u/d [September 1918], 12, box 909, ZGN, RG 45, NA.
10. Lovett's report is located in box 963, ZPA, RG 45, NA, and is excerpted in Paine, *First Yale Unit*, vol. II, 172–4.
11. A paraphrase of President Wilson's statements in 1917. Lovett remarks, Yale Unit reunion transcript (1924), 140, Davison Papers, YA.
12. Wortman, *Millionaires' Unit*, 203.
13. Klachko, *Admiral William Shepherd Benson: The First Chief of Naval Operations*, 94–5.
14. Sims, *The Victory at Sea*, 253.
15. U.S. Navy, Office of Naval Records and Library, *The American Naval Planning Section London*, 91–116.
16. Lord, "History of United States Naval Aviation, 1898–1939," 453–5, 456–7.
17. See H. A. Jones, *The War in the Air*, vol. 6, 381–382. When operational demands shifted RAF air assets away from Dunkirk, "there were expectations of important help from America." See Jones, *The War in the Air*, vol. 6, 383–4.
18. Letter, Cone to Sims, March 1, 1918, box 52, Sims Papers, LC.
19. Paris Headquarters Log, March 11 and 14, 1918.
20. Arthur, *Contact!*, 13; Paine, *First Yale Unit*, vol. I, 98–101; Rossano, *Price of Honor*, 114.
21. In a separate conversation with Commandante de Patrouille Loubigniac, Ens. G. A. Smith and Lt. Harry Guggenheim learned they could expect no advanced bombers from France in 1918. Memo, Lovett to Cone, March 19, 1918, box 1, Entry 35 [Belgian Bombing to France, Misc.], RG 72, NA; Headquarters Log, March 18, 1918, box 1, GA-1, RG 45, NA; Memo, Guggenheim/Smith to Cone, March 22, 1918, SS/Ga 144, SHM.
22. Wortman, *Millionaires' Unit*, 201; Letter, Lovett to Adele Brown, quoted in Greer, *The Millionaires' Unit: U.S. Naval Aviators in the First World War*, documentary film script; Paris Headquarters Log, March 20, 1918.
23. Lovett personal correspondence quoted in Greer, documentary script.

24. Rossano, *Price of Honor*, 123; McDonnell's descriptions of his experiences is contained in Paine, *First Yale Unit*, vol. II, 181–5.

25. Aartick is located about ten miles southwest of Bruges. See O'Connor, *Airfields and Airmen*, 63–7.

26. A description of this raid is provided in Peter Wright, "Dunkerque Days and Nights"; see also Paris Headquarters Log, March 28 and April 1, 1918. McDonnell's flights on board high-altitude day bombers exposed him to the extreme discomforts often encountered on such missions. Ken MacLeish, one of the Yale flyers, a former student of McDonnell's then serving with nearby 213 Squadron, offered a complete catalogue of ailments afflicting pilots and observers shoehorned into tight cockpits without heat or oxygen. In early April, he experienced frozen fingers, crushing headaches, extreme nausea, weakness and dizziness, and ringing in his ears. "The veins in my ears expanded enormously at every heartbeat and cut off my hearing entirely," he reported. MacLeish became so impaired he could hear neither the roar of his engine nor the sharp crack of his machine guns. Rossano, *Price of Honor*, 134.

27. Rossano, *Price of Honor*, 114; Wortman, *Millionaires' Unit*, 205.

28. Lovett's description of these bombing raids is contained in Paine, *First Yale Unit*, vol. II, 176–8, and Wortman, *Millionaire's Unit*, 205–7.

29. Wortman, *Millionaires' Unit*, 206–7; Lovett correspondence quoted in Greer, *The Millionaires' Unit: U. S. Naval Aviators in the First World War*. Bewsher, *Green Balls: The Adventures of a Night Bomber*; "The Dawn Patrol," in *A Treasury of War Poems*, ed. Clarke; *The Bombing of Bruges*.

30. Bewsher, *Green Balls*; "The Dawn Patrol"; *The Bombing of Bruges*.

31. Cone also offered surplus personnel to the British during the ongoing crisis. Cone was on his return from London to Paris, see Paris Headquarters Log, March 28, 1918.

32. Wortman, *Millionaires' Unit*, 209–10; see also Wright, "Dunkerque Days and Nights," 138. Poet Brewsher survived the crash and later memorialized Allan in one of his works. See also H. A. Jones, *The War in the Air*, vol. 6, 386–7.

33. Lovett's report is reprinted in Paine, *First Yale Unit*, vol. II, 175–80.

34. Ibid.

35. McDonnell, "Report on Observation of British Day and Night Bombing," March 29, 1917, box 1, Entry 35 [Belgian Bombing to France, Misc.], RG 72, NA. See also Paine, *First Yale Unit*, vol. II, 181–5.

Chapter 8. The Great Debate

1. Outgoing cable #1320, Sims to OpNav, March 18, 1918, in NBG Summary, June 24, 1918, box 910, ZGU, RG 45; also box 1, Belgian Bombing to France, Misc. (Entry 35), RG 72, both NA.

2. Outgoing cable #1355, March 23, 1918, in NBG Summary, June 24, 1918, box 910, ZGU, RG 45, NA. Cone's response contained in Sims' dispatch #5496 of March 23, 1918, in box 6, Entry 23, RG 72, NA.

3. Sims dispatch #5496 of March 23, 1918, in Correspondence between the Department and Vice Admiral Sims, relative to the Northern Bombing Squadron, u/d [September 1918], box 6, Entry 23, RG 72, NA.

4. The brooding specter of the War Department and the Navy's fear of offending Army aviators is discussed in Lord, "History of United States Naval Aviation, 1898–1939," 457–60.

5. OpNav to Sims #4676, April 4, 1918, box 6, Entry 23, RG 72; incoming cable #1882, Sims to Cone, April 9, 1918, box 910, ZGU, RG 45, NA.

6. Pink quoted in Memo, Cone to Sims, April 17 and 18, 1917, box 3, World War I Collection, Aviation History Unit, NHHC; also Cone to Sims, Memo re: Dunkirk Bombing Project, May 11, 1918, box 910, ZGU, RG 45, NA. Note: in the April 1, 1918, amalgamation of the RFC and RNAS, Pink was designated Staff Officer 1st Class—Marine Operations, Directorate of Flying Operations

7. Cone to Sims, April 17, 1918, box 910, ZGU, RG 45, NA; Cone to Sims, April 18, 1918, AIR 1/70/15/9/122, NAUK.

8. U.S. Naval Aviation Forces, Foreign Service (USNAFFS) was the formal designation of Cone's command. Letter, Edwards to Cone, April 25, 1918, and Cable #6960, Sims to OpNav, April 27, 1918, both box 6, Entry 23, RG 72, NA.

9. In his response, Cone noted that some of the General Board's recommendations were based on outdated information. He received authorization (incoming cable #1971, April 20, 1918) to proceed three days later. Delays sending, receiving, decoding, and forwarding cables between Washington, London, and Paris caused considerable confusion, with various requests, approvals, and counterproposals bypassing each other. This definitely affected proposals to secure Caproni bombers, whether in the United States, from Italy, or from the U.S. Army. Note: there is some confusion here. Department cabled April 18, 1918, that conditions indicated impossibility of obtaining Capronis in Italy; should attempt to secure planes from Army/Pershing per incoming cable 1963; but Army had already indicated no planes available, see incoming cable 1650, April 23, 1918. The cables must have crossed.

10. B. L. Smith to Cunningham, Confidential memorandum, u/d, box 1, Cunningham Collection, MCL.

11. See Northern Bombing Group Summary, 2.

12. Memo to Secretary of the Navy, April 30, 1918, box 910, ZGU, RG 45, NA; also Letter, Cunningham to Geiger, April 30, 1918, box 1, Cunningham Collection, MCL.

13. Letter, Cunningham to Geiger, April 30, 1918, box 1, Cunningham Collection, MCL; Meeting Notes, "Proposed bombing project in the Dunkirk area," April 29, 1918, box 3, World War I Collection, Aviation History Unit, NHHC. A cable to Sims May 1 indicated training was being arranged in conformance with this agreement. See Northern Bombing Group Summary, 2, 19; see also OpNav dispatch #5622 to Sims, May 1, 1918, in Correspondence between the Department and Vice Admiral Sims, relative to the Northern Bombing Squadron, u/d [September 1918], box 6, Entry 23, RG 72, NA.

14. This conclusion had been drawn from minutes of a conference held May 4, 1918, and recommendations regarding the 1919 program dated May 3.

15. Daniels to Sims #6682, May 28, 1918, and OpNav dispatch #6632 to Sims, same date, both in Correspondence between the Department and Vice Admiral Sims, relative to the Northern Bombing Squadron, u/d [September 1918], box 6, Entry 23, RG 72, NA; also Sims to Cone, May 31, 1918, box 910, ZGU, RG 45, NA; the Executive Committee Log of June 6, 1918, mentions that telegram #2449 dated June 4, 1918, with reference to NBS, was discussed, "Cone to make reply today."

16. Memo, McDonnell to Benson, June 2, 1918, box 3, World War I Collection, Aviation History Unit, NHHC.

17. Cable #2085, Cone to Sims, June 3, 1918, box 3, World War I Collection, Aviation History Unit, NHHC.

18. Sims to OpNav #9180, June 8, 1918, in Correspondence between the Department and Vice Admiral Sims, relative to the Northern Bombing Squadron, u/d [September 1918], box 6, Entry 23, RG 72 NA; Sims to Daniels, June 12, 1918, box 3, World War I Collection, Aviation History Unit, NHHC.

19. Irwin's response described in Northern Bombing Group Summary, 3. See also Letter, Cunningham to Geiger, June 1, 1918, box 1, Cunningham Collection, MCL.

20. Letter, Cone to Irwin, August 17, 1918, quoted in Northern Bombing Group Summary, 4. At the end of the summer, Captain Knox and Captain Yarnell of the London Planning Section described "a coastal patrol of very fast land machines . . . carried out daily by British flying machines. . . . We are inclined to think that this type of land machine . . . will take the place of seaplanes for coastal patrol." Ibid.

21. Northern Bombing Group Summary, 5.

22. Lord, "History of Naval Aviation, 1898–1939," 461–3. See also Lord, Naval Aviation History (draft), 171.

23. Both the Aircraft Production Board recommendation and subsequent War Department approval are quoted in Northern Bombing Group Summary, 5. See Daniels to Baker, March 7, 1918.

24. Army-Navy correspondence regarding aircraft to be provided to naval aviation is summarized in several documents. See, for example, Daniels–Baker Correspondence, box 910, ZGU, RG 45, NA; also Northern Bombing Group Summary, 5–7, and NBG Correspondence Summary, box 6, Entry 23, RG72, NA.
25. Daniels to Baker, Confidential, April 4, 1918, in NBG Correspondence Summary, box 6, Entry 23, RG 72, NA.
26. Letters and cables recorded in NBG Summary, box 909, ZGN, RG 45, NA. A Department summary of NBG correspondence compiled at the end of the summer 1918 points out possible confusion between Sims and Washington, the Department envisioning attacks *against submarines in their bases* and Sims advocating *attacks against the bases themselves.*
27. War Department, Memo for Chief of Staff, April 10, 1918, and Memo, Baker to Daniels, Confidential, April 10, 1918, signed Benedict Crowell. Also, Daniels to Baker, April 15, 1918, saying Sims had been advised to contact Pershing and that Navy did desire Army to arrange supply of seventy-five two-place fighters. Daniels–Baker Correspondence, box 910, ZGU, RG 45, NA.
28. Baker to Daniels, May 2, 1918; Daniels to Baker, May 14, 1918; also Daniels to Baker, May 4, 1918. Efforts to allocate critical but scarce Liberty engines generated even more heated disputes.
29. Letter, Irwin to Benson, June 5, 1918, quoted in Lord, "History of United States Naval Aviation, 1898–1939," 521.

Chapter 9. Night Bombers Needed
1. Bolling cable No. 70, July 31, 1917, Cablegrams to and from France, 1917–1919, Entry 96, RG 18, NA.
2. A. B. Gregg, "History of Caproni Biplane (Night Bombing)," Bureau of Aircraft Production History, File 452.1, Entry 22, RG 18, NA-CP (hereafter BAPH).
3. Bolling cable No. 70.
4. Gregg, "History of Caproni Biplane," 2.
5. Caproni wanted one million dollars down and another million when the first airplane crossed the Atlantic. Waldon to Potter, July 10, 1918, BAPH. The Caproni biplane was an improved version of the Caproni triplane that could carry almost the same useful load, but was less complicated to manufacture.
6. See "Great Battleplane New Idea in Aviation," *New York Times,* August 5, 1917, and "The Giant Airplane War's New Demand," *New York Times,* July 22, 1917. Source: *New York Times* Online Archive.
7. "Aeroplane Board Visit to Northwest," *Flying* 6, no. 8 (September 1917), 674.
8. Holley, *Ideas and Weapons,* 44, 184.

9. Ibid., 43.
10. Joint Army-Navy Technical Aircraft Board to SecWar and SecNav, May 29, 1917, as reproduced in Mauer, *U.S. Air Service in World War I*, vol. 2, 105.
11. Pershing cable No. 73S, August 12, 1917, Caproni, Cablegrams to and from France, Entry 96, RG 18, NA-CP (hereafter Caproni Cablegrams).
12. Pershing Cable No. 96S, August 12, 1917, Caproni Cablegrams.
13. Report of Army and Navy Technical Members of the Aeronautical Commission sent to Europe, September 4, 1917, File 542.1, Bureau of Aircraft Production, General. Correspondence 1917–1919, Entry 22, RG 1, NA-CP (hereafter BAP-GC); see also Lord, "History of United States Naval Aviation 1898–1939," 356.
14. Gregg, "History of Caproni Biplane," 3.
15. Clark, Memorandum for CSO of the Army, September 12, 1917, 11, BAP-GC.
16. Squier to Pershing, Cable No. 172R, September 14, 1917, Caproni Cablegrams.
17. Bruce, "Handley Page 0/100 and 0/400," *Flight*, February 27, 1953, 259.
18. "10,000 Planes to Fly to France," 815; "America Urged to Build Bombing Planes," *Air Service Journal*, August 23, 1917.
19. Gregg, "History of Handley Page Night Bombing Airplanes," 1, BAPH.
20. Ibid., 3.
21. Bolling Cable No. 196, October 9, 1917, Caproni Cablegrams.
22. Bevione to Potter, September 10, 1918, "Summary of the Negotiations with the American Aviation for the Construction of the Caproni Machines (hereafter Summary of Negotiations), 1, File 542.1 Caproni, BAP-GC.
23. Gregg, "History of Caproni Biplane," 4–5.
24. According to Gregg, Bolling met with Caproni and Italian government representatives on November 30, 1917. See "History of Caproni Biplane," 5.
25. Congress regularized the Aircraft Production Board's authority in October 1917, redesignating it the Aircraft Board.
26. Waldon Memorandum for Potter, July 10, 1918, 2, File 542.1 Caproni, BAP-GC.
27. Summary of the Negotiations, 3.
28. Waldon to Potter, September 19, 1918, 1, File 542.1 Caproni, BAP-GC.
29. Ibid.
30. Summary of Negotiations, 4; Waldon Memorandum for Mr. Potter, 2. Note: Waldon gave the Italian representative as "Monsuer Gracci," without specifying first name or title, but this was undoubtedly Giuseppi Grassi as listed in numerous publications of the era (e.g., *Flight and the Aircraft Engineer*, vol. 10, part 2, 22).
31. Waldon Memorandum for Potter, July 10, 1918, 3.

32. U.S. Congress, Aircraft Production Hearings Before the Subcommittee of the Committee on Military Affairs United States Senate, 65 Cong., 2d sess., vol. 1, 406–7, 1918.

33. Gregg, "History of Caproni Biplane," 7.

34. Frank Brisco, Report of Caproni Situation in U.S.A., September 4, 1918, 1, File 542.1, BAP-GC.

35. Gregg, "History of Caproni Biplane," 6–8.

36. Gregg, "History of Handley Page Night Bombing Airplanes," 4.

37. Ibid., 5–6.

38. Sweetser, *The American Air Service*, 203.

39. Pershing Cable No. 283, December 16, 1917, Extracts of Cablegrams to and from France, Entry 22, RG 18, NA.

40. Gregg, "History of Handley Page Night Bombing Airplanes," 7.

41. Foulois cable of January 26, 1918, as cited by Gregg, "History of Handley Page Night Bombing Airplanes," 7.

42. Sweetser, *The American Air Service*, 204.

43. Gregg, "History of Handley Page Night Bombing Airplanes," 8.

44. Ibid., 10.

45. Maj. Gen. Mason M. Patrick, "Final Report of Chief of the Air Service A.E.F. to the Commander in Chief American Expeditionary Forces," *Air Service Information Circular* II, no. 180 (February 15, 1921): 40.

Chapter 10. Putting the Plan into Motion

1. The numbers of personnel assigned to headquarters duty varied greatly week to week.

2. Memorandum, Cone to OpNav, April 1, 1918; Executive Committee Log, April 3 and May 8, 1918; Letter, Sims to Daniels, April 8, 1918, World War I Collection, Aviation History Unit, NHHC.

3. Letter, Cone to Irwin, April 1, 1918, cc: Bureaus of Navigation, Construction, Yards, Steam Engineering, Ordnance. It carried the endorsement of Admiral Sims to Secretary Daniels. World War I Collection, Aviation History Unit, NHHC. See also, Letter, Cone to Irwin, May 11, 1918, box 156, GU, RG 45, and Irwin Weekly Conferences, box 64, Aviation Section Reports 1918–1919 (Entry 36), RG 72, NA.

4. These American-built versions of the Airco DH-4 originally designed by Geoffrey DeHavilland were modified to utilize the Liberty engine developed in the United States in 1917 and were sometimes referred to as Liberty DH-4s.

5. "McDonnell has just returned from Dayton and says DHs are wonderful," in Letter, Cunningham to Geiger, June 1, 1918, box1, Cunningham Collection, MCL. See also Minutes, Irwin weekly conferences, April 24 and May 1, 1918.

6. Letter, Mims to Cunningham, June 26, 1918, box 1, Cunningham Papers, MCL.

7. Grey was now in the RAF, having transferred into that organization with the rank of lieutenant colonel (later changed to wing commander) when the RNAS and RFC were amalgamated on April 1, 1918. It was originally proposed that the RAF ranks were to be derived from existing Royal Navy and Army ranks, but this did not come to pass and on August 1, 1919, Air Ministry Weekly Order 973 introduced new rank titles for RAF officers. This has led to much confusion regarding the ranks of RAF officers held between April 1 and August 1, 1918, especially for those who had been in the RNAS—such as Grey—who held both naval and naval aviation ranks.

8. Lord, Naval Aviation History (draft), 175.

9. Memo, Briscoe to Cone, April 17, 1918; Executive Committee Log, April 27, 1918.

10. Executive Committee Log, May 7 and 8, 1918. Right from the beginning Cone and the rest of the staff recognized that all scheduled operations were completely dependent on timely arrival of promised aircraft. See Memorandum re: Dunkerque Bombing Project, May 21, 1918, box 910, ZGU, RG 45, NA.

11. Executive Committee Log, June 26, 1918.

12. See Edwards to Cone, May 27, 1918, box 910, ZGU, RG 45, NA.

13. Lambe to Cone, May 24, 1918, AIR 1/70/15/9/122, NAUK.

14. Sims to Secretary of the Admiralty, June 12, 1918, box 910, ZGU, RG 45, NA.

15. Cone to Edwards, June 26, box 910, ZGU, RG 45, NA; Cone to Edwards, June 29, 1918, box 145, GN, RG 45, NA; Cables Nos. 2124 and 2471, both June 6, 1918, in NBG Summary. See also Headquarters Log, June 28, and Executive Committee Log, June 29, 1918. Additional correspondence regarding this "mix-up" is located in box 145, GN, and box 910, ZGU, both RG 45, NA. The NBG was soon placed under the command of Vice Admiral, Dover. See Lambe to Cone, July 1, 1918, box 145, GN, RG 45, NA.

16. Craven, "History of U.S. Naval Aviation, French Unit," 35–51, box 912, ZGU, and Progress Report, USNAFFS, March 15, 1918, box 131, GA-1, both RG 45, NA.

17. Letter, Cone to Irwin, May 11, 1918, box 156, GU, RG 45, NA. "I have just returned from three days spent in the Dunkirk area, which incidentally, we have officially named the 'Northern Bombing Squadrons.' " The Night Bombing Squadrons were eventually designated the Night Bombing Wing, commanded after mid-August by Robert Lovett and the Day Bombing squadrons were designated the Day Bombing Wing, led by Marine aviator Alfred Cunningham.

18. Executive Committee Log, April 25, 1918; Sims dispatch #2051 to OpNav, July 30, 1918, recommended change of name to Northern Bombing Group. Reply of August 2, OpNav dispatch #9401 to Sims, authorized change, both in Correspondence between the Department and Vice Admiral Sims, relative to the Northern Bombing Squadron, u/d [September 1918], box 6, Entry 23, RG 72, NA; see also Irwin Weekly Conference minutes, August 6, 1918.

19. Letter, MacLeish to Priscilla Murdock, July 28, 1918, in Rossano, *Price of Honor*, 193. That this was more than youthful hero worship is evidenced by Lovett's later business career at Brown Brothers Harriman and government service as Assistant Secretary of War for Air during World War II, Undersecretary of State, and finally, Secretary of Defense during the Korean War.

20. Memorandum re: Dunkerque Bombing Project, May 13, 1918, box 910, ZGU, RG 45, and Letter, Cone to Irwin, May 11, 1918, box 156, GU, RG 45, both NA.

21. Executive Committee Log, May 2, 1918.

22. In a memorandum dated May 13, Cone reported, "Commander D. C. Hanrahan is the future Commanding Officer." See Memorandum re: Dunkerque Bombing Project, May 13, 1918, box 910, ZGU, RG 45, NA. See also Headquarters Log, May 15, 17, 20, and Executive Committee Log, May 14, 1918.

23. Rossano, *Stalking the U-boat*, 322, and *The Price of Honor*, 215.

24. Headquarters Log, May 18, 27, and 28, 1918, and Executive Committee Log, May 31, 1918.

25. Headquarters Log June 4, 18, 26, 27, and July 17, 1918.

26. Hanrahan, Progress Report, 9–10.

27. Letter, Cunningham to Long, August 10, 1918, box 2, Cunningham Collection, MCL. As late as October 20, 1918, Cunningham reported, "My relations with Captain Hanrahan and the other officers at Group Headquarters are very friendly and there is not the slightest indication of friction." In Cunningham to Long, October 20, 1918, box 2, Cunningham Collection, MCL.

28. Letter, Cunningham to Long, September 4, 1918, box 2, Cunningham Collection, MCL. Grey was now a lieutenant colonel in the RAF, which would soon be changed to wing commander though he would retire as a commander in the Royal Navy.

29. Letter, Cunningham to Long, September 14, 1918, box 2, Cunningham Collection, MCL. Far down the chain of command, Lt. (jg) Ken MacLeish, then serving at Eastleigh, wrote to his fiancée [September 15, 1918], "The old Iron Duke . . . he's always on the go. He hasn't any idea what he's doing,

but he's always doing something, and he always gets balled up." Rossano, *Price of Honor*, 215.

30. Letter, MacLeish to Priscilla Murdock, July 28, 1918, in *Price of Honor*, 193.

31. As early as August 1917, Naval Constructor George Westervelt had suggested to Admiral Sims that several officers of the RNAS "especially experienced in administrative affairs" be attached to Captain Cone's staff. Westervelt to Sims, August 18, 1918, quoted in Lord, "History of United States Naval Aviation, 1898–1939," 372.

32. Executive Committee Log, May 9, 1918.

33. Ibid., July 17, 1918. In late July, Admiral Sims reported the imminent arrival in the United States of Captain Ashworth, RAF, "for duty training personnel for Northern Bombing Squadron."

34. "I have arranged for a British expert in bombing and formation flying to come to Miami for two weeks to help you out. Get everything he knows before he leaves." Letter, Cunningham to Geiger, June 6, 1918, box 1, Cunningham Collection, MCL. See also Sims dispatch #1871 to OpNav, July 27, 1918, in Correspondence between the Department and Vice Admiral Sims, relative to the Northern Bombing Squadron, u/d [September 1918], box 6, Entry 23, RG 72, NA.

35. Progress Report, July 15, 1918.

36. See Cone, Memorandum re: Dunkerque Bombing Project, May 13, 1918, and Memorandum, Cone to Sims, May 22, 1918, box 910, ZGU, RG 45, NA.

37. All three eventually received the award, though not without some controversy. See Cone to Sims, September 8, 1919, and Sims to Daniels, September 8, 1919, in box 52, Sims Papers, LC.

38. Headquarters Log, June 26, 1918; Executive Committee Log, July 2, 17, 1918. The Navy experience mirrored, on a smaller scale, the dependence of the U.S. Air Service on the support of the British, French, and Italian air forces. See particularly Sebastian Cox, "Aspects of Anglo-US Cooperation in the Air in the First World War," 27–33; and Rossano, "Doing Their Duty Side by Side: Allied–American Aviation Cooperation in World War I" (master's thesis, UNC–Chapel Hill, 1974).

Chapter 11. Training of Personnel

1. Progress Report, July 15, 1918, box 133, GA-1, RG 45, NA. For disagreements over allocation of personnel for various naval aviation initiatives see Rossano, *Stalking the U-boat*, 328.

2. Letter, F. P. Evans to Edwards, May 11, 1918, box 133, GA-2, RG 45, NA.

3. Letter, Cone to Irwin, May 11, 1918, box 156, GU, RG 45, NA.

4. Letter, Edwards to Cone, May 5, 1918, and Letter, Edwards to Davies, May 14, 1918, both box 133, GA-2, RG 45, NA. Edwards also suggested that the

Navy operate its own school for day and night bombing at Killingholme, a suggestion never implemented.

5. Training would include two weeks of elementary instruction and then two weeks of actual night flying. Headquarters Log, April 29, May 20, and June 4, 1918; Executive Committee Log, May 1, 6, 15, 16, and 21, 1918. See also Memorandum re: Dunkerque Bombing Project, May 21, 1918, box 910, ZGU, RG 45, NA.

6. Pilots included Charles Basset, David Ingalls, Kenneth MacLeish, David Judd, Woldemar Crosscup, Edwin Pou, Arthur Boorse, Charles Beach, Charles Hodges, Albert Johnson, Ralph Loomis, Thomas McKinnon, George Moseley, Horace Schermerhorn, Earl Schoonmaker, Malcom West, Cornelius O'Connor, and Archibald McIlwaine.

7. All pilot biographies in Arthur, *CONTACT!*

8. Moseley, *War Letters*, 166–7.

9. Possibly Cone and A.W.K Billings. See Headquarters Log.

10. Rossano, *Price of Honor*, 157; Rossano, *Hero of the Angry Sky*, 170; Sheely and Sheely, *Sailor of the Air*, 131.

11. Moseley, *War Letters*, 180–1. See also Hudson, *History of the American Air Service in World War I*, 39–40, and Maurer, ed., *The United States Air Service in World War I*, vol. III, 58, 97, 108.

12. Head of Ordnance, Instruments, and Accessories Section in Intelligence and Planning Division, Paris headquarters. Upon his return, Donnelly recommended the Navy acquire a DH aircraft for use in fitting bombing gear. Headquarters Log, June 11 and 18, 1918.

13. See Progress report, July 15, 1918. See also numerous references in Paine, *First Yale Unit*, vol. II; Sheely and Sheely, *Sailor of the Air*; Moseley, *War Letters*; Rossano, *Price of Honor* and *Hero of the Angry Sky*. Other information and photographs are contained in the Randall Browne Collection, Photographic Unit, NHHC, and Irving Sheely Collection, privately held.

14. Progress Report, July 18, 1918, box 133, GA-1, RG 45, NA.

15. Executive Committee Log, May 9, 1918.

16. Headquarters Log, April 10, June 17, and 18, 1918; Executive Committee Log, May 9, 1918.

17. These were the Airco-manufactured DH-9s designed by Geoffrey DeHavilland and powered with 240-hp Siddeley Puma engines. Naval and Marine aviators serving with RAF squadrons in the summer/fall of 1918 flew this aircraft.

18. Colin Owers, "Handley Page Trainee Pilot: The Experiences of Sir Laurence Hartnett in the RNAS and RAF," 115. See also Owers, "Handley Page 0/400," 120–30.

19. Owers, "Trainee Pilot," 115–16.

Chapter 12. Capronis Coveted

1. OpNav dispatch #4676, April 4, 1918, "Correspondence Between the Department and Vice Admiral Sims, Relative to the Northern Bombing Group," 1–2, attached to "Summary," Classified Correspondence Relating to Aviation 1917–1919, Entry 23, RG 72, NA.
2. Sims dispatch #6834, April 21, 1918, 2 [see above].
3. U.S. Navy, Naval Aviation History Office, Procurement of Aviation Material in Europe 1917–1918, 3.
4. Paine, *First Yale Unit*, vol. II, 214.
5. Versus 85 mph for the Ca. 3.
6. Paine, *First Yale Unit*, vol. II, 215.
7. Rossano, *Stalking the U-boat*, 304. When the RFC was absorbed into the newly established RAF on April 1, 1918, Grey was assigned to that service for the duration of the war.
8. Ibid., 44, 47.
9. Ibid., 288.
10. U.S. Navy, Naval Aviation History Office, "Procurement of Aviation Material in Europe 1917–1918" (hereafter "Procurement of Aviation Material"), 4–7.
11. Cable for the Chief of Staff and Chief Signal Officer, April 12, 1918, reproduced in Army Reorganization Hearings before the Committee on Military Affairs, 66th Cong., 1st sess. (hereafter Army Reorganization Hearings), 951.
12. Foulois testimony, Army Reorganization Hearings, 952–3.
13. Cable for Chief of Staff, May 3, 1918, reproduced in Army Reorganization Hearings, 952.
14. Cable No. 1275-R, May 9, 1918, reproduced in Army Reorganization Hearings, 952–3.
15. Shirley, "La Guardia, Caproni Bombers, and the U.S. Navy," 128.
16. Maurer, "Flying with Fiorello," 113–14.
17. Shirley, "John Lansing Callan," 183.
18. Shirley, "La Guardia, Caproni Bombers, and the U.S. Navy," 129.
19. Note: Foulois' postwar testimony alleging that "in August, 1917, the Air Service American Expeditionary Forces had placed orders with the Italian government for a large number of Caproni bombing planes," appears to be a gross exaggeration of the facts. He seems to have confused the activities of the Bolling Mission with the large orders later placed for construction in the United States.
20. Callan to Cone, March 27, 1918, box 3, Callan Papers, LC.
21. Callan to Cone, March 28, 1918, box 3, Callan Papers, LC.
22. Train's complaints are outlined in a May 29, 1918, letter from Cone to Irwin, box 3, Callan Papers, LC.

23. Shirley, "LaGuardia, Caproni Bombers, and the U.S. Navy," 131. See also Irwin Weekly Conference, May 18, 1918, box 64, Aviation Section Reports 1918–1919 (Entry 36), RG 72, NA; and Lord, "History of United States Naval Aviation, 1898–1939," 252.

24. "Procurement of Aviation Material," 20.

25. Letter, Cone to Irwin, May 29, 1918, box 3, Callan Papers, LC.

26. Sims to Cone (Incoming Cable 2246), May 15, 1918, box 910, ZGU, RG 45; see also Rossano, *Stalking the U-boat*, 306.

27. Cone to Sims (Outgoing Cable 1891), May 16, 1918, box 910, ZGU, RG 45, NA.

28. Memo, Patrick to Royal Italian Aeronautical Mission, Paris, May 31, 1918, box 3, Callan Papers, LC. See also Shirley, "Capt. La Guardia," 131–2.

29. Bludworth, *Battle of Eastleigh, England USNAF*, 30; Potter, "The 'Fiat' and the 'Caproni,'" 96–8.

30. Paine, *First Yale Unit*, vol. II, 219–22.

31. Callan to Lovett, June 15, 1919; Callan to Cone, July 16, 1918, both box 3, Callan Papers, LC; Arthur, *CONTACT!*, 5.

32. Walker to Callan, June 29, 1918; Callan to Walker, July 5, 1918; Callan to Cone, July 16, 1918; La Guardia to Patrick, July 21, 1918; Callan's response to Cone, August 12, 1918, all box 3, Callan Papers, LC.

33. Callan to Cone, July 12, 1918, box 3, Callan Papers, LC.

Chapter 13. Airbases and Support Facilities

1. Hanrahan, Progress Report, 1–2.

2. Headquarters Log, April 15, 1918; Executive Committee Log, April 19, 1918.

3. OpNav dispatch #6757 to Sims, May 31, 1918, and Sims dispatch #9408 to OpNav, June 2, 1918, in Correspondence between the Department and Vice Admiral Sims, relative to the Northern Bombing Squadron, u/d [September 1918], box 6, Entry 23, RG 72, NA. See also Hanrahan, Progress Report, 2.

4. Letter, Cone to Irwin, June 5, 1918, quoted in Lord, Naval Aviation History (draft), 179.

5. Headquarters Log and Executive Committee Log, both June 18, 1918. On Lambe's promotion, see "Air Vice-Marshal Sir Charles Lambe," *Air of Authority—A History of RAF Organisation*, Royal Air Force Organisation website (http://www.rafweb.org/Biographies/Lambe.htm0); also "Aircraft in the Zeebrugge and Ostend Battles," *Flight*, February 27, 1919, 268.

6. Headquarters Log and Executive Committee Log, both July 8, 1918; Progress Report, July 15, 1918. A Sims dispatch of July 6 observed, "For night bombing squadron will probably establish advance repair base at Eastleigh as one Base B." Correspondence between the Department and Vice Admiral Sims, relative to the Northern Bombing Squadron,

u/d [September 1918], box 6, Entry 23, RG 72, NA; see also Hanrahan, Progress Report, 2–3.

7. Sims dispatch #1232 to OpNav, July 16, 1918, in Correspondence between the Department and Vice Admiral Sims, relative to the Northern Bombing Squadron, u/d [September 1918], box 6, Entry 23, RG 72, NA.

8. Letter, Cone to Irwin, July 19, 1918, box 156, GU, RG 45, NA.

9. OpNav dispatch #OL 198 to Sims, August 21, 1918, reported, "All remaining material for Repair Base B and Northern Bombing Group will be shipped to Southampton instead of Pauillac." Correspondence between the Department and Vice Admiral Sims, relative to the Northern Bombing Squadron, u/d [September 1918], box 6, Entry 23, RG 72, NA.

10. These would be the Liberty-engined DH-4s manufactured in the United States. See Headquarters Log, July 9 and 17, 1918, and Executive Committee Log, July 31, 1918. A communication from Washington dated July 17 foretold difficulties, however, reporting that "materials, provisions, as well as materials and small stores for Northern Bombing Group have been shipped to Pauillac. 28 DH planes shipped to Pauillac . . . practically all personnel for Navy squadrons for Northern Bombing Group have been transferred to Pauillac." OpNav dispatch #8665 to Sims, July 17, 1918, in Correspondence between the Department and Vice Admiral Sims, relative to the Northern Bombing Squadron, u/d [September 1918], box 6, Entry 23, RG 72, NA.

11. Headquarters Log, April 12, May 3, and 9, 1918, and Executive Committee Log, April 17, 1918.

12. Executive Committee Log, May 19, and Headquarters Log, June 19, 1918.

13. Lord, Naval Aviation History (draft), 178.

14. Lord, Naval Aviation (History), 179; Hanrahan, Progress Report of Northern Bombing Group from July 1 to November 11, 1918, 3–4, in box 3, Cunningham Collection, MCL, hereafter Hanrahan, Progress Report. See also Progress Report, NBG update, August 15, 1918, box 133, GA-1, RG 45, NA.

15. An account of the devastating June raids that drove RAF bombers [214 Squadron] away from Coudekerque is contained in Wright, "Dunkerque Days and Nights," 140–1. See also, Progress Report, July 15, 1918, box 133, GA-1, RG 45, NA.

16. Headquarters Log, June 23, July 6, 12, 16, and 17, 1918; Progress Report, NBG update, August 15, 1918. Observers sent to St. Inglevert: Horton Tatman, Howard Clarke, August Cook, Louis Fallows, Jay Jones, George Lowry, Clarence Meyers, Matthew O'Gorman, Charles Rourke, George Sprague, Virgil Stevens, Cyrus Sylvester, Alva Turnbull, Earl Van Calder, Anthony White. See also, Sheely and Sheely, Sailor of the Air, 157–9.

17. Progress Report, July 18, 1918.

18. Ibid.; Campagne station log, RG 24, NA; Hanrahan, Progress Report, 4.
19. Progress Report, July 18, 1918; Hanrahan, Progress Report, 4.
20. From a distance, Lt. Harvey Mims, the Marine aviation officer at Washington headquarters, wrote in late June, "I have just learned by reading over the secret and confidential files in operations (aviation), the exact location of the squadron and I must say it is a wonderful location. It is level and hard ground for miles and miles and the squadron numbers will be 7—8—9—10. Think it best not to mention the name of the town by mail." Letter, Mims to Cunningham, June 28, 1918, box 1, Cunningham Collection, MCL. Executive Committee Log, July 29, 1918. Hanrahan reports on situation that developed regarding cutting of crops at Oye field.
21. Letter, Ingalls to Father, July 22, 1918, in Rossano, *Hero of the Angry Sky*, 205.
22. Letter, Cunningham to Geiger, May 11, 1918, box 1, Cunningham Collection, MCL; Headquarters Log, June 5 and 10, 1918; Arthur, *CONTACT!*, 56. Chamberlain later became embroiled in a sharp controversy over whether his exploits reported in the newspapers and elsewhere had ever actually occurred. See, for example, Cunningham to Long, October 5, 1918. "The more we look into Chamberlain's affair, the worse it looks. . . . It looks as if Chamberlain manufactured the whole story in Paris and gave it to the newspaper himself." Box 2, Cunningham Collection, MCL.
23. Headquarters Log, July 15, 1918.
24. Johnson, *Marine Corps Aviation*, 19; Arthur, *CONTACT!*, 9–10.

Chapter 14. Send in the Marines
1. Useful summaries of Marine Corps aviation activities in the 1912–18 period can be found in Edward Johnson, *Marine Corps Aviation: The Early Years 1912–1940*; Robert Sherrod, *History of Marine Corps Aviation in World War II*, 2nd ed.; Roger Emmons, *Marine Corps Aviation in World War I: Flying Personnel*; and Peter Mersky, *U.S. Marine Corps Aviation Since 1912*, 4th ed. A definitive account of Marine Corps aviation during World War I remains to be written.
2. Of this group, six were detached for duty in the Azores and twelve sent to Hazelhurst Field in New York. Karl Day interview, MCL.
3. Francis Mulcahy interview, MCL.
4. Letter, Day to Parents, October 6, 1917, Day Papers, MCL. This is the trip described in Cunningham's published diary.
5. Kelly's predictions were not too far off the mark. The first two groups (when combined) traveled eventually to Europe, the coast patrol men moved on to NAS Cape May (New Jersey), and the balloon men went to Omaha. Letter, Kelly to Mother, October 9, 1917, Kelly Papers, MCL.
6. See Mulcahy interview, MCL.

7. Stoff, *The Aviation Heritage of Long Island.*

8. Letter, Day to Father, October 19, 1917, Day Papers; Letter, Kelly to Mother, February 20, 1918, Kelly Papers, both MCL.

9. See Richards interview, *Over The Front,* Autumn 1992, 196. [AU: where is this in Biblio?]

10. Day interview and u/d letter, and Mulcahy interview, both MCL; James Ralph Doolittle was the thirty-seventh volunteer of the Lafayette Escadrille. On May 2, 1917, while flying SPAD at G.D.E. he was injured severely when it side-slipped into the ground. After recuperation, he was released from French aviation. He returned to America and became a civilian instructor for the U.S. Air Service. On July 26, 1918, while flying at Kenilworth Field near Buffalo, New York, he crashed from an altitude of three hundred feet, killing his passenger and fatally injuring himself.

11. Letter, Henry to Cunningham, November 23, 1917, box 1, Cunningham Collection, MCL.

12. Day interview and Mulcahy interview, both MCL; December 1917 was the coldest month on record in New York, below zero almost every night.

13. Letter, Kelly to Mother, u/d, Kelly Papers; Gray, Letter to Father, January 11, 1918, Gray Papers, both MCL.

14. William R. Evinger, *Directory of Military Bases in the U.S.,* 47; The Southwest Louisiana Historical Association, www.swlahistory.org/Gerstner Field.

15. Letter, Kelly to Mother, January 25, 1918, Kelly Papers, MCL.

16. Letter, Day to Father, January 11, 1918; Gray interview; Letter, Kelly to Mother, January 20, 1918, Kelly Papers, all MCL.

17. Letter, Kelly to Mother, January 20, 1918, Kelly Papers, MCL.

18. Day interview; Mulcahy's log book contains entry, "instruction on stick," in Mulcahy Papers, both MCL.

19. Letter, Kelly to Mother, January 15, 1918, Kelly Papers, MCL.

20. Letter, Kelly to Mother, March 10, 1918, Kelly Papers, MCL.

21. Letters, Kelly to Mother, February 6 and 10, 1918, Kelly Papers, MCL.

22. Letter, Kelly to Mother, February 20, 1918, Kelly Papers, MCL.

23. Letters, Kelly to Parents, February 5, 26, 28, and March 3, 1918, Kelly Papers, MCL.

24. Letter, Kelly to Mother, u/d, Kelly Papers, MCL.

25. Letters, Kelly to Mother, February 2, 6, and March 26, 1918, Kelly Papers, MCL.

26. Letter, Cunningham to Major General Commandant, February 6, 1918, in box 3, Cunningham Collection, MCL.

27. "Have taken over two JN-4-Bs, Bennett letting us use all spares we need, B wishes to use remaining Jennys until he qualifies his students, a week or two, and then Marines will get them." Geiger to Cunningham, March 15, 1918, box 1, Cunningham Collection, MCL.

28. Letters, Cunningham to Geiger, March 23 and 26, 1918, and Geiger to Cunningham, March 30 and April 9, 1918, all box 1, Cunningham Collection, MCL.

29. Letters, Roben to Cunningham, April 22, 1918, and Cunningham to Geiger, May 28 and June 1, 1918, box 1, Cunningham Collection, MCL.

30. Letters, Roben to Cunningham, April 17, 1918, and Mims to Cunningham, June 28, 1918, box 1, Cunningham Collection, MCL.

31. Letter, Mims to Cunningham, June 24, 1918, box 1, Cunningham Collection, MCL.

32. Letters, Kelly to Mother, April 10, 1918, Kelly Papers; Geiger to Cunningham, May 8, 1918, box 1, Cunningham Collection; Darden interview, all MCL.

33. Richards interview, MCL; Letters, Kelly to Mother, June 19 and April 13, 1918, Kelly Papers, MCL.

34. Ford Rogers interview, quoted in Johnson, *Marine Corps Aviation: The Early Years*, 17.

35. Darden interview, MCL.

36. Adam Wait and Noel Shirley, " 'Devil Dog' Sam Richards"; Richards interview, MCL; Letter, Cunningham to Geiger, May 3, 1918, box 1, Cunningham Collection, MCL, requests info regarding naval officers who had volunteered for Marine service, wants to process disenrollments while he was stationed in Washington.

37. Telegram, Roben to Cunningham, April 17, and Letters, Roben to Cunningham, April 17 and 18, 1918, all box 1, Cunningham Collection, MCL. Smith had served as assistant naval attaché in Paris throughout Hall's time in France and may have heard some of the stories that swirled around him.

38. Letters, Cunningham to Geiger, May 7, 11, and June 11, 1918; Day interview, all MCL.

39. Pilots spent a day or two at NAS Miami. "We flew everything they had, N-9, R-6, and a flying boat they called an F-boat." Mulcahy interview, MCL.

40. Letter, Talbot to Mother, April 29, 1918, Talbot Papers; Day interview, Day Papers, both MCL.

41. Darden and Richards interviews, both MCL.

42. Ibid.

43. "Got hold of a DH or two, got some familiarization in those." Mulcahy interview, MCL.

44. Day interview, MCL.

45. 2nd Lt. Melville Sullivan, NA# 596, died May 7, 1918, in a crash at Marine Flying Field (Miami). See Arthur, *CONTACT!*, 185. Letters, Mims to Cunningham, June 21, 1918, and Cunningham to Geiger May 11, 1918, both box 1, Cunningham Collection, MCL.

46. Day interview, MCL.

47. Mulcahy interview, MCL.
48. Letter, Day to Family, May 7, 1918, Day Papers, MCL.
49. Letters, Geiger to Cunningham, April 2, 1918, and Mims to Barnett, July 8, 1918, both box 1, Cunningham Collection, MCL.
50. Letter, Day to Family, May 7, 1918, Day Papers MCL; Richards interview, both MCL.
51. Letters, Kelly to Mother, April 10, 22, and 28, 1918, Kelly Papers, MCL.
52. "Saw a fine movie 'The Face in the Dark.' " Letter, Kelly to Mother, June 3, 1918, Kelly Papers, MCL.
53. Richards interview; Letters, Kelly to Mother, April 10, 17, 22, and June 19, 1918, Kelly Papers, all MCL.
54. According to Sam Richards, soon after Cunningham's arrival in Miami the men started calling him "Ma" because he tended to be somewhat sissy and always followed orders to the letter. Richards interview, *Over the Front*, Autumn 1992.
55. Squadrons A, B, C, and D were later redesignated Squadrons 7, 8, 9, and 10 for duty in France. Ultimately, Marine squadrons became the Day Wing/NBG, with Major Cunningham as commanding officer.
56. Letter, Day to Parents, June 21, 1918, Day Papers, MCL.
57. The old Curtiss Field, now a busy complex renamed Marine Flying Field, continued operations, completing training for Squadron D.
58. Letter, Day to Mother, July 14, 1918, Day Papers, MCL.
59. The USS *DeKalb* was the former *Prinz Eitel Friedrich*, a German raider voluntarily interned at Norfolk in 1915. When the United States entered World War I, the vessel was seized by Customs officials and transferred to the Navy. Reconditioned and refitted as a troop transport, the ship was renamed USS *DeKalb*.
60. Johnson, *Marine Corps Aviation*, 19.
61. Richards interview, *Over the Front*, Autumn 1992.
62. Variations include Slippery Sam or Polish Red Dog.
63. "While disembarking, a sling loaded with foot lockers broke, spilled into harbor, soaked everything." Diary of John Benson, in newsletter of First Marine Aviation Force Veterans Newspaper, u/d, in Karl Day Papers, MCL.
64. Letters, Cunningham to Long, July 31, 1918, and Cunningham to Mims, August 4, 1918, both box 2, Cunningham Collection, MCL. Admiral Sims reported the arrival of the Marines in dispatch #2545 to OpNav, August 7, 1918, in Correspondence between the Department and Vice Admiral Sims, relative to the Northern Bombing Squadron, u/d (September 1918), box 6, Entry 23, RG 72, NA.
65. Letter, Day to Mother, August 7, 1918, Day Papers, MCL.

66. A memo dated June 30, 1918, quoted in Northern Bombing Group Summary, u/d (September 1918), listed fifty-two DH4s shipped abroad as of that date, ninety-three held in Philadelphia awaiting shipment, six en route from the factory, and four at the Miami flying field, box 909, ZGN, RG 45, NA.

67. The Airco DH-9 aircraft, when modified to accept the American-manufactured Liberty Engine was redesigned as the DH-9A. "Among the British aircraft supplied to the U.S. Navy for use by the Northern Bombing Group in France in 1918 were 54 D.H. 9A two-seat bombers powered by 400 hp American Liberty engines. These were not assigned U.S. Navy serial numbers, but flew with American roundels being painted over the British on the wings and fuselage roundel being painted out and replaced by station markings." Swanborough and Bowers, *United States Naval Aircraft since 1911*, 544.

68. All aircraft equipping Marine Corps day bombing squadrons consisted of either Liberty-engined DH-4s manufactured in the United States or Liberty-engined DH-9As obtained from the British.

69. Letter, Day to Mother, August 7, 1918, Day Papers, MCL.

70. Richards interview, *Over the Front*, Autumn 1992.

Chapter 15. Learning from the British

1. Headquarters Log, June 18, 27, and July 6, 1918.

2. See Hanrahan, Progress Report, 12. During this period Americans served with 213, 214, 217, and 218 squadrons, RAF.

3. Others later served with 213, 214, and 217 squadrons.

4. The 82nd Wing consisted of 38 Squadron, a night bombing unit equipped with FE-2b aircraft; 214 Squadron, a night bombing group flying Handley Pages; and 218 Squadron.

5. Quoted in Sheely and Sheely, *Sailor of the Air*, 149.

6. Rossano, *Price of Honor*, 187–8; also Sheely and Sheely, *Sailor of the Air*, 147–51, which includes extracts from MacLeish's pilot reports.

7. Moseley, *War Letters*, 188.

8. Paine, *First Yale Unit*, vol. II, 251–2.

9. Moseley, *War Letters*, 189–91, 194–5.

10. Paine, *First Yale Unit*, vol. II, 252; Rossano, *Hero of the Angry Sky*, 200.

11. The British Women's Army Auxiliary Corps formed in 1917.

12. Rossano, *Hero of the Angry Sky*, 200–1, 203.

13. The compact town of Bergues, once an important port, retained most of its early fortifications. Sometimes referred to as "the other Bruges of Flanders," it did not become part of France until 1668.

14. Rossano, *Price of Honor*, 177.

15. Arthur, *CONTACT!*, 48.

16. McIlwaine's adventures are chronicled in Paine, *First Yale Unit*, vol. II, 255–9.

17. Memo, Gates to Hanrahan, August 22, 1918; see also Memo, Toulon to Gates, August 20, 1918; Telegram, Headquarters, 5th Group, RAF to Hanrahan, August 21, 1918, all box 144, GN, RG 45, NA. Other Navy pilots serving with 217 Squadron included Albert R. "Babe" Johnson of Moorefield, West Virginia, who trained at Bay Shore, Key West, and Moutchic. Arthur, *CONTACT!*, 78.

18. See Wright, "Dunkerque Days and Nights," 131, 140–1.

19. Cacutt, ed., *Classics of the Air: An Illustrated History of the Development of Military Planes from 1913–1935*, 17–24. See also *Jane's Fighting Aircraft of World War I*, 72–4, and Owers, "Handley-Page 0/400," 120–30.

20. Rossano, *Stalking the U-boat*, 334.

21. A memo from the NBG Operations Officer to Captain Hanrahan dated October 1, 1918, identified seven pilots and aerial gunners serving with 214 Squadron, including Taylor, William Gaston, and Harry P. Davison, another of the Yale boys. The Army Air Service and Hugh Trenchard's Independent Force (IF) operated a similar program. Ultimately, thirty-six Americans carried out missions with the IF; eighteen were killed, wounded, or captured. See Hudson, *Hostile Skies*, 250.

22. Hanrahan, Progress Report, November 11, 1918, 12–13 box 3, Cunningham Collection, MCL. Hanrahan's report contained a very complete recapitulation of Northern Bombing Group activities from July 1 until the end of the war.

Chapter 16. Operating against the Enemy: The Campaign Begins

1. Confidential Bulletin #12, July 27, 1918, and Cable, Coombe to Callan, July 28, 1918, both box 156, GU, RG 45, NA; Paine, *First Yale Unit*, vol. II, 223; Arthur, *CONTACT!*, 98, 125.

2. The flights across the Alps began on August 19. McDonnell took off from Taliedo and flew to Turin from which he attempted to cross the Alps.

3. Paine, *First Yale Unit*, vol. II, 216–17, 224, 232–41. McDonnell to Hanrahan, and Train to Cone, both August 12, 1918, and McDonnell to Cone, August 13, 1918. Complete recapitulation of the saga of the Navy Capronis in Memo, Hanrahan to Cone, August 24, 1918; see also Aviation Ministry to Turin, August 20, 1918, all correspondence in box 156, GU, RG 45, NA.

4. "Since August 15 no war flights have been made with Caproni planes. Work has been constantly carried on with the view of rendering these planes fit for service, but until the cessation of hostilities, the only war activities carried out by the Night Wing were those of the personnel attached to Number 214 Squadron, R.A.F." Hanrahan, Progress Report, 13–14.

5. St. Inglevert Station Log, September 16, 1918, RG 24, NA.
6. Record of Proceedings of a Board of Inquest . . . in the case of Clyde Palmer and Phillip Frothingham," September 15, 1918; and Memo, Lovett to Hanrahan, September 15, 1918, both in box 910, ZGU, RG 45, NA. See also unsigned, undated report, box 461, PA, RG 45, NA, and Paine, *First Yale Unit*, vol. II, 238–9.
7. St. Inglevert Station Log, September 2 and November 1, 1918; Arthur, *CONTACT!*, 53, 43; Willis and Carmichael, *United States Navy Wings of Gold*, 194.
8. St. Inglevert Station Log, September 2, 1918. On August 24, Grey had crashed Caproni B-4 at Orly field in Paris "due to tied controls. No casualties." He had previously ferried Caproni B-5 across the Alps in early August, the B-5 being the same aircraft that carried out the NBG's only successful raid against Ostend on August 15.
9. St. Inglevert Station Log, September 3, 1918.
10. St. Inglevert Station Log, October 14 and 26, 1918. See also NBG Signals Log, October 13 and 14, 1918.
11. Campagne Station Log, October 24, 28, and 30, 1918.
12. Campagne Station Log, September–November 1918, passim. See also Letter, Edwards to Hanrahan, October 31, 1918, box 133, GA-2, RG 45, NA.
13. The Varssenaere raid involved six squadrons and inflicted heavy damage. See H. A. Jones, *The War in the Air*, vol. 6, 395–6.
14. Ingalls' career with 213 Squadron is detailed in Rossano, *Hero of the Angry Sky*, 215–77.
15. Ibid., 292.
16. This was the day the Allied offensive in Flanders resumed in earnest, leading to great demands on aviation units as all squadrons carried out offensive missions across the lines.
17. For MacLeish's short career with 213 Squadron and his death, see Rossano, *The Price of Honor*, 221–36, and Paine, *First Yale Unit*, vol. II, 350–73.
18. The activities of the NAS Dunkirk pilots who flew with the French is recounted in Rossano, *Stalking the U-boat*, 76–7, and Paine, *First Yale Unit*, vol. II, 313–28. See also Moseley, *War Letters*, 207–17.

Chapter 17. Operating against the Enemy: Day Bombing
1. Rossano, *Price of Honor*, 197.
2. Ibid., 202–3.
3. Chevalier to Cone, August 16, 1918, quoted in Lord, Naval Aviation History [draft], 112.
4. The history of the repair and assembly repair facility at Eastleigh is recounted in detail in Bludworth, *The Battle of Eastleigh, England, USNAF*. See especially 9–13, 86–90.

5. Charles Todd, "A Brief History of the Work Accomplished by Squadron 'C' First Marine Aviation Force. In France," box 2, World War I Aviation Papers, MCL. Karl Day later flew five similar missions with the British and dropped 1,000 pounds of bombs.

6. Audembert is located southwest of Calais, near St. Inglevert. See NBG Signal Log, September 4, 1918.

7. "This policy was made possible through courtesy of the 5th Group, RAF." Hanrahan, Progress Report, 14. The personnel detailed to 217 Squadron on August 21 included Benjamin Harper, Clyde Bates, Frank Nelms, Chapin Barr, T. S. Cravatte, and Herbert Hughes. From untitled excerpts FMAF Daily Log, box 3, Cunningham Collection, MCL. According to an NBG Operations memo dated October 1, 1918, five Marine crews were in action with 218 Squadron on September 28 and 29. See also Weekly Report, 8th Squadron, October 25, 1918, in box 2, World War I Aviation Papers, MCL. See roster compiled by Roger Emmons, box 1, World War I Aviation Papers, MCL.

8. A DH-4, Navy A-3295. This was part of a group of 155 aircraft obtained through agreements with the U.S. Army negotiated in the spring of 1918.

9. This included subjecting engines to a two-hour block test, a test flight incorporating tailspins and vertical banks, machine-gun synchronization tests, and a thorough tightening of all wires and connections. See Memo, Cunningham to Hanrahan, October 16, 1918, box 2, World War I Aviation Papers, MCL.

10. Edwards to Cone, August 7, 1918, box 133, GA-2, RG 45, NA.

11. Regarding these discussions, see Letters, Edwards to Cone, August 7 and 10, 1918, both in box 133, GA-2, RG 45, NA.

12. A comparison of the performance of DH-9 and DH-9A aircraft is contained in H. A. Jones, *The War in the Air*, vol. 6, 545–6. The painfully slow buildup of serviceable aircraft can be traced in the various Weekly Reports and Daily Progress Reports filed by squadron and wing commanding officers, found in box 2, World War I Aviation Papers, MCL. See also John Morrow, "Defeat of the German and Austro-Hungarian Air Forces in the Great War, 1909–1918," 125, in Higham and Harris, eds., *Why Air Forces Fail: The Anatomy of Defeat*; also, NBG Signal Log, passim, September–October 1918.

13. NBG Signal Log, September 28, 1918, discussed the battlefield situation on the opening day of the advance.

14. Chapin C. Barr, Navy Cross Citation, *Military Times Hall of Valor*, http://projects.militarytimes.com/citations-medals-awards/.

15. Both men were awarded the Navy Cross for this action.

16. Frank Nelms Jr., Distinguished Service Medal Citation, *Home of Heroes*, http://www.homeofheroes.com.
17. Documented in NBG Signal Log, September 30, 1918.
18. H. A. Jones, *The War in the Air*, vol. 6, 534–5.
19. See Sherrod, *History of Marine Corps Aviation in World War II*; also Todd, "A Brief Outline," 3–4. During World War I, the Distinguished Service Medal "outranked" the Navy Cross.
20. According to A Memorandum for Director of Naval Aviation dated October 10, 1918, two American DH-4s had been attached to 215 Squadron and five Navy pilots were serving with 218 Squadron, box 910, ZGU, RG 45, NA.
21. Hanrahan, Progress Report, 14–15; Johnson, *Marine Corps Aviation*, 24; Todd, "A Brief Outline," 3. Beginning on the evening of October 3, RAF units returned to Lichtervelde repeatedly, inflicting considerable damage. H. A. Jones, *The War in the Air*, vol. 6, 535.
22. Letter, Cunningham to Long, October 5, 1918, Cunningham Collection, Box 2, MCL. See also Memo, Wemp to Hanrahan, October 6, 1918, box 2, World War I Aviation Papers, MCL, and Johnson, *Marine Corps Aviation*, 20–5.
23. Day interview, MCL.
24. The group had already been hit by influenza, "which played havoc with their personnel," and was subsequently quarantined, but likely spread the disease to the other Marine units. See Letter, Cunningham to Long, October 12, 1918, box 2, Cunningham Collection, MCL.
25. Letter, Cunningham to Long, October 20, 1918, box 2, Cunningham Collection, MCL.
26. Extracted from Confidential Bulletin #229 from Admiral Sims, November 1, 1918; see also E. B. Taylor and A. H. Wright, Pilot Raid Reports, October 14, 1918, in box 2, World War I Aviation Papers, MCL. Pilots unable to reach the targets typically dropped their ordnance in the sea before returning to base. See also NBG Signal Log, October 14, 1918. This raid was part of a massed assault on German communications and transportation systems carried out by aircraft dropping forty tons of bombs. H. A. Jones, *The War in the Air*, vol. 6, 539–40.
27. These exploits are recounted in several sources, including Johnson, *Marine Corps Aviation*, 24; see also Todd, "A Brief Outline," 6, and especially Roger Emmons, "The Raid on Thielt: First Marine Air Combat Operation," in box 2, World War I Aviation Papers, MCL. See also Talbot/Robinson Medal of Honor citations, and NBG Signal Log, October 14, 1918. The initial communiqué reported Robinson as "probably fatally wounded."

28. See NBG Signal Log, October 14, 1918. First reports transcribed at 2 p.m. simply stated that Lytle and Wiman were missing, with their machine, D-3, "last seen near Westrude travelling northwest."

29. Pilot Raid Reports document poor flying conditions encountered in October with notations like "visibility very poor," "cloudy and rain," "poor," "impossible to reach objective on account of low clouds," "unable to judge the [bomb] hits due to heavy fog or clouds." Reports contained in box 1, World War I Aviation Papers, MCL. See also Johnson, *Marine Corps Aviation*, 24–5.

30. Marine aviators carried out two raids on October 17 against Steenbrugge railway junction, the rail line between Steenbrugge and Eecloo, and Zeebrugge mole, dropping 6,156 pounds of bombs. See Memorandum for Chief of Staff—Aviation, October 18, 1918, box 910, ZGU, RG 45, NA.

31. Operations Orders #3 and #4, October 16 and 17, 1918, both in box 2, World War I Aviation Papers, MCL.

32. See, for example, 9th Squadron Raid Order No. 5, October 17, 1918, and 7th Squadron Raid Order No. #1, October 17, 1918, both in box 2, World War I Aviation Papers, MCL.

33. American aviators had much to say about the Liberty-powered DH-4s. According to Maj. George Reinburg, commanding officer of the 2nd Day Bombardment Group, the aircraft suffered from numerous deficiencies, including poor performance at high altitude with full loads, limited bomb load, dangerously positioned and operated fuel tank, poorly configured observer's cockpit, weak tail structure and undercarriage, and over-heavy and powerful engine. See Maurer, ed., *The U.S Air Service in World War I*, vol. IV, 87–8.

34. See NBG Signal Log, October 19, 1918.

35. Todd Diary, reprinted in "A Brief Outline," 8–9.

36. Cablegram, Hanrahan to Sims, October 25, 1918. See especially Colgate Darden interview, 30–3, copy at MCL of 1984 USNI interview, and Todd, "A Brief Outline," 9.

37. See Memorandum for Chief of Staff—Aviation, October 18, 1918, box 910, ZGU, RG 45, NA, and Extract from Confidential Bulletin #231, November 7, 1918, Benson Papers, LC.

38. The problem was later discovered to be ignition trouble, with the generator discharging until the motor stopped.

39. Statement made by 2nd Lt. John Frederick Gibbs, November 5, 1918, and similar statement by 2nd Lt. Frank Nelms Jr., same date. See also Memo, Nelms and Gibbs to Cunningham, November 22, 1918, in box 1, World War I Aviation Papers, MCL, and NBG Signal Log, October 27, 1918. Record keeping at the time claimed four enemy aircraft destroyed and one more

shot down out of control. Given the confusion of battle, conflicting claims, and the fact that many of the encounters occurred far behind enemy lines, the exact tally will likely never be known. Marine losses were calculated at one plane/crew lost in combat and a second downed by anti-aircraft fire. That aircraft glided into Holland where authorities seized the plane and interned the crew until the Armistice.

40. Todd, "A Brief Outline," 2–3.
41. Hanrahan, Progress Report, 17
42. Todd, "A Brief Outline," 12; Letter, Cunningham to Long, October 31, 1918, box 2, Cunningham Collection, MCL
43. Cunningham to Long, October 31 and November 9, 1918, box 2, Cunningham Collection, MCL.

Chapter 18. Bombing, Bombing, and More Bombing

1. This document is contained in Northern Bombing Group Summary, u/d [September 1918], box 909, ZGN, RG 45, NA.
2. Quoted in Lord, Naval Aviation History (draft), 187.
3. Edwards to Cone, August 5, 1918, quoted in Lord, "History of Naval Aviation, 1898–1939," 558–9.
4. Arthur, CONTACT!, 41–2.
5. Letter, Ames to Davison, September 29, 1918, Davison Papers, YA.
6. See Callan to Cone, August 14 and September 5, 1918, quoted in Lord, "History of United States Naval Aviation, 1898–1939," 568, 571. See also "Heads in Agreement to the Constitution of the Inter-Allied Independent Air Force," October 3, 1918, in H. A. Jones, The War in the Air, vol. 7, 41. This document identified the objectives of the Independent Air Force as attacks on German industry, commerce, and population.
7. Sims Cable #3545, August 25, 1918, and Cable #IL 1446, September 12, 1918, both Correspondence between the Department and Vice Admiral Sims, relative to the Northern Bombing Squadron, u/d [September 1918], box 6, Entry 23, RG 72, NA. See also Rossano, Stalking the U-boat, 74–6.
8. See Guggenheim to Cone, September 11, 1918, and Memo (Edwards?) to Cone, September 12, 1918, both quoted in Lord, "History of United States Naval Aviation, 1898–1939," 566–7. See also Lord, Naval Aviation History (draft), 114.
9. Sims to Benson, "Highly secret," September 25, 1918, quoted in Lord, "History of United States Naval Aviation, 1898–1939," 568. See also Sims to Daniels, September 26, 1918, box 3, Belgian Bombing to France, Misc. (Entry 35), RG 45, NA, and Rossano, Stalking the U-boat, 343–4, 350–1.
10. Letter, Cone to Irwin, September 24, 1918, quoted in Lord, "History of United States Naval Aviation, 1898–1939," 570.

11. Rossano, *Stalking the U-boat*, 339. See also Bludworth, *Battle of Eastleigh, England, USNAF*, 96–8.

12. See AEF/USAS Technical section, Memo, Wolf to Burton, September 1, 1918, file 542.1, Entry 22, RG 18, NA. See also Henry Lockhart Jr., Special Mission, Bureau of Aircraft Production, October 9, 1918. Before the war, Lockhart served as president of the Simplex Motor Car Company and the Wright Aeroplane Company. See also "Caproni Biplane (Night Bombing)," Records of the Air Service, Bureau of Aircraft Production History 1917–1926, box 3, Entry 22, RG 18, NA, and Caproni Documents re: U.S. Production, file 542.1, Entry 22, RG 18, NA.

13. Letters, Edwards to Cone, June 3 and July 25, 1918, box 133, GA-2, RG 45, NA. Operational V/1500s first entered service in October 1918.

14. Edwards to Cone, August 7, 1918, box 133, GA-2, RG 45, NA. Cone referred obliquely to these machines in his September 24 missive to Irwin.

15. For the development of this program see Lord, Naval Aviation History (draft), 159–61. For Eaton Interview see Manuscript Collection, Archives Branch, NHHC.

16. "This idea was taken up in a large way by Admiral Benson." Lord, Naval Aviation History (draft), 103.

17. Events at NAS Killingholme are documented in Owers, "Killingholme Diary," 169–95, 211–32. See also Rossano, *Stalking the U-boat*, 169–85.

18. Memo, Sims to Cone, August 4, 1918 [August 10], box 463, PA, RG 45, NA.

19. Edwards to Cone, August 7, 1918, box 145, GP, RG 45, NA.

20. See Lord, "History of United States Naval Aviation, 1898–1939," 342–4, and Turnbull and Lord, *History of Naval Aviation*, 135–6.

21. Morton, *Mustin—A Naval Family of the Twentieth Century*, 113–14.

22. Quoted in ibid., 114. According to Morton, CNO Benson had approved the concept.

23. Ibid., 115.

24. Efforts to obtain a Caproni from the Italian Mission in the United States, or from Hugo d'Annunzio, representative of the Caproni factory, proved equally fruitless.

25. Letter, Taylor to Potter, July 13, 1917, and Memo, Re: Informal Conference, July 17, 1917, both File 542, Entry 22, RG 18, NA.

26. Letter, Daniels to Baker, u/d file copy, File 542, Entry 22, RG 18, NA. This correspondence also came with the approval of CNO Benson, indicated by his initials on the file copy of the document.

27. Bellinger memoir, box 1, Bellinger Papers, Archives Branch, NHHC; Morton, *Mustin*, 116. The N-9 launch occurred March 7, 1919, with pilot Frank Johnson at the controls. A second takeoff, this time with an observer on board, took place the next day.

28. History Memoir of USNAF in Italy, box 919, ZGU, RG 45, NA. All cables, box 919, ZGU, RG 45, NA.
29. These events are described in detail in Rossano, *Stalking the U-boat*, 287, 302, 312, 348, 351–2. See also box 3, Callan Papers, LC.
30. Both of whom compiled long and distinguished service records.
31. Letter, Cone to Irwin, September 24, 1918, box 156, GU, RG 45, NA. Irwin eventually did travel to Europe to inspect the naval aviation effort, but by that time Cone had been injured seriously during the sinking of RMS *Leinster* by an enemy submarine. When the two met, they definitely did not get along. See Rossano, *Stalking the U-boat*, 356, 357–8.
32. Letter, Edwards to Cone, October 23, 1918, in box 133, GA-2, RG 45, NA.

Chapter 19. Lessons and Legacies

1. A total of 1,090 individuals served with the First Marine Aviation Force: 164 Marine officers, 846 Marine enlisted personnel, 14 Navy officers, and 66 Navy enlisted personnel. See box 1, World War I Aviation Papers, MCL.
2. In *The Wise Men: Six Friends and the World They Made*, Walter Isaacson and Evan Thomas refer to Lovett's "beloved bomber forces," 195, 202–7. See especially 205–6. The U.S. Bombing Survey carried out by the Air Service shortly after World War I stressed the necessity of bombing a specific industrial target and not a town generally; determining scientifically the particular industry and factory to be attacked most likely to affect overall production; bombing a single target day and night on several successive days to completely destroy it; and eschewing bombing for morale effect. See Maurer, *The United States Air Service in World War I*, vol. IV, 501–3.

Selected Bibliography

Archival Collections

Franklin D. Roosevelt Presidential Library and Museum (FDRPLM), Hyde Park, N.Y.

Library of Congress (LC), Manuscript Division, Washington, D.C.
  Mitchell, William, Papers of

Marine Corps Library (MCL) and Archives, Quantico, Va.

National Archives of the United Kingdom (NAUK), Kew, England

National Archives of the United States (NA), Washington, D.C.

National Archives of the United States (NA-CP), College Park, Md.
  Records of the American Expeditionary Forces (World War I), 1917–23.
    Gorrell's History of the American Expeditionary Forces Air Service, 1917–19, microfilm series M990.
  Records of the Army Air Forces, RG 18

Naval History and Heritage Command (NHHC), Washington, D.C.
  Aviation History Division
    Northern Bombing Group Subject File
  Archives Branch
    Lord, Clifford L., Papers of

Service Historique de la Marine (SHM), Paris

Yale University Library and Archives (YA), New Haven, Conn.

Public Documents

U.S. Congress, House of Representatives. Army Reorganization. Hearings before the Committee on Military Affairs, 66th Cong., 1st sess., Washington, D.C., 1920.

———. Hearings Before the Subcommittee No. 1 (Aviation) on the Select Committee on Expenditures in the War Department, 66th Cong., *War Expenditures*, vol. 3. Washington, D.C.: GPO, 1920.

———. Hearings: Naval Investigation, 66th Cong., Washington, D.C., 1921, 1849–51.

U.S. Congress, Senate. Aircraft Production Hearings Before the Subcommittee of the Committee on Military Affairs United States Senate, 65 Cong., 2d sess., vol. 1. Washington, D.C.: GPO, 1918.

Books and Articles
"10,000 Planes to Fly to France." *Aerial Age Weekly* 7, no. 12 (July 8, 1918): 815–16.
Abbatiello, John. *Anti-Submarine Warfare in World War I: British Naval Aviation and the Defeat of the Submarine*. London: Routledge, 2006.
"Aeroplane Board Visit to Northwest." *Flying* 6, no. 8 (September 1917): 674.
Albion, Robert G. *Makers of Naval Policy, 1798–1947*. Annapolis, Md.: Naval Institute Press, 1981.
Alexander, Sir Walter A. *The War in the Air*, vol. 1. Oxford: Clarendon Press, 1922.
"America Urged to Build Bombing Planes." *Air Service Journal*, August 23, 1917, 210.
"Appropriation for Aeronautics." *Air Service Journal*, July 12, 1917, 13.
Arthur, Reginald Wright. *CONTACT! Careers of Naval Aviators Assigned Numbers 1–2,000*. Washington, D.C.: Naval Aviation Register, 1967.
Ash, Eric. *Sir Frederick Sykes and the Air Revolution, 1912–1918*. London: Routledge, 1999.
Atkinson, J. L. Boon. "Italian Influence on the Origins of the American Concept of Strategic Bombardment." *Air Power Historian* 4, no. 3 (July 1957): 141–9.
Bacon, Sir Reginald. *The Dover Patrol*. New York: G. D. Doran, 1919.
Bartlett, Charles P. *In the Teeth of the Wind: The Story of a Naval Pilot on the Western Front, 1916–1918*. Annapolis, Md.: Naval Institute Press, 1994.
Bewsher, Paul. *The Bombing of Bruges*. London: Hodder and Stoughton, 1918.
———. *Green Balls: The Adventures of a Night Bomber*. Edinburgh: William Blackwood and Sons, 1919.
Bludworth, T. Francis. *The Battle of Eastleigh, England, USNAF*. New York: Thompson, 1919.
Bowers, Peter M. *Curtiss Aircraft 1907–1947*. London: Putnam, 1979.
Boyle, Andrew. *Trenchard*. London: Collins, 1962.
Bradford, James. "Henry T. Mayo: Last of the Independent Naval Diplomats." In James Bradford, ed., *Admirals of the New Steel Navy*. Annapolis, Md.: Naval Institute Press, 2012.
Bruce, J. M. "The De Havilland D.H.9." Historic Military Aircraft No. 12, Part 1. *Flight*, November 8, 1956, 385–7, 392.
———. "Handley Page 0/100 and 0/400." Historic Military Aircraft No. 4. *Flight*, February 27, 1953, 254–9.

———. "The Sopwith 1½ Strutter." *Flight*, September 28, 1956, 545.

———. "The Sopwith Tabloid, Schneider and Baby." Historic Aircraft No. 17, Part 1. *Flight*, November 8, 1957, 733–6.

Cacutt, Lew, ed. *Classics of the Air: An Illustrated History of the Development of Military Planes from 1913–1935*. New York: Exeter Books, 1988.

Castle, Ian. *London 1917–18: The Bomber Blitz*. Oxford: Osprey, 2010.

———. *The Zeppelin Base Raids—Germany 1914*. Oxford: Osprey, 2011.

Clarke, George H., ed. *A Treasury of War Poems*. Boston: Houghton Mifflin, 1917.

Coffman, Edward. *The War to End All Wars*. Madison: University of Wisconsin Press, 1986.

Cooke, James C. *Billy Mitchell*. Boulder, Colo.: Lynne Rienner, 2002.

Cooper, Malcolm. "The British Experience of Strategic Bombing." *Cross & Cockade International* 17, no. 2 (Summer 1986): 49–61.

Copp, Dewitt S. *A Few Great Captains: The Men and Events That Shaped the Development of U.S. Air Power*. McLean Va.: EPM Publications, 1980.

Cosmas, Graham A., ed. *Marine Flyer in France: The Diary of Captain Alfred A. Cunningham, November 1917–January 1918*. History and Museums Division, Headquarters U.S. Marine Corps, Washington, D.C., 1974.

Cox, Sebastian. "Aspects of Anglo-US Cooperation in the Air in the First World War." *Air and Space Power Journal* 18, no. 4 (Winter 2004): 27–33.

Cronon, E. David. *The Cabinet Diaries of Josephus Daniels, 1913–1921*. Lincoln: University of Nebraska Press, 1963.

Culver, Edith Dodd. *The Day the Air-Mail Began*. Kansas City, Kan.: Cub Flyers, 1971.

———. *Tailspins: A Story of Early Aviation Days*. Santa Fe, N.M.: Sunstone Press, 1986.

Daniels, Josephus. *Our Navy at War*. New York: Doran, 1922.

———. *The Wilson Era: Years of War and After, 1917–1923*. Chapel Hill: University of North Carolina Press, 1946.

Daugherty, Leo J. *Pioneers of Amphibious Warfare, 1898–1945: Profiles of Fourteen American Military Strategists*. Jefferson, N.C.: McFarland, 2009.

Dodds, R. V. "Britain's First Strategic Bombing Force." *The Roundel* 15, no. 6 (July–August 1963). http://www.manitobamilitaryaviationmuseum.com/ PDF/No. 3 Wing.pdf.

Emmons, Roger. *Marine Corps Aviation in World War I: Flying Personnel*. Glenview, Il.: Marine Corps Aviation Association, 1973.

Evinger, William R. *Directory of Military Bases in the U.S.* Phoenix: Oryx Press, 1991.

Fairlie, John A. *British War Administration*. London: Oxford University Press, 1919.

Fisher, William E. *The Development of Military Night Aviation of 1919*. Maxwell AFB, AL: Air University Press, 1998.

"Forgotten Trench Diggers of the Western Front—Meet WW1's Chinese Labour Corps." *Military History Now*, December 4, 2013. http://militaryhistorynow

.com/2013/12/04/the-forgotten-trench-diggers-of-the-western-front-meet-ww1s-chinese-labour-corps/.

Foulois, Benjamin D., with C. V. Glines. *From the Wright Brothers to the Astronauts: The Memoirs of Major General Benjamin D. Foulois*. New York: McGraw-Hill, 1968.

Freidel, Frank. *Franklin D. Roosevelt: The Apprenticeship*. Boston: Little, Brown, 1952.

Futrell, Robert F. *Ideas, Concepts, Doctrine: Basic Thinking in the United States Air Force 1907–1960*. Maxwell Air Force Base, Ala.: Air University Press, 1989.

Gilbert, Martin. *Churchill: A Life*. New York: Holt, 1991.

———. *The First World War: A Complete History*. New York: Henry Holt and Co., 1994.

Gorrell, Edgar. "An American Proposal for Strategic Bombing in World War I." *Air Power Historian* 5, no. 2 (April 1958): 102–17.

Greer, Thomas H. "Air Arm Doctrinal Roots, 1917–1918." *Military Affairs* 20, no. 4 (Winter 1956): 202–16.

———. *The Development of Air Doctrine in the Army Air Arm 1917–1941*. Washington, D.C.: Office of Air Force History, 1985.

Hedin, Robert. *The Zeppelin Reader*. Iowa City: University of Iowa Press, 1998.

Higham, Robin, and Stephen Harris, eds. *Why Air Forces Fail: The Anatomy of Defeat*. Lexington: University Press of Kentucky, 2006.

Hildreth, Alonzo. "Over There—World War I." *All Hands* (June 1962): 56–63.

Holley, I. B., Jr. *Ideas and Weapons: Exploitation of the Aerial Weapon by the United States During Word War I; a Study in the Relationship of Technical Advance, Military Doctrine, and the Development of Weapons*. New Haven, Conn.: Yale University Press, 1953.

Hudson, James. *Hostile Skies, A Combat History of the American Air Service in World War I*. Syracuse: Syracuse University Press, 1968.

Hunt, Andrew W. "From Production to Operations, the US Aircraft Industry: 1916–1918." In *2003 Logistics Dimensions*, pp. 191–208.

Hurley, Alfred F. *Billy Mitchell: Crusader for Air Power*. Bloomington: Indiana University Press, 1975.

Isaacson, Walter, and Evan Thomas. *The Wise Men: Six Friends and the World They Made*. New York: Simon and Schuster, 1986.

Jane's Fighting Aircraft of World War I. New York: Military Press, 1990.

Johnson, Edward C. *Marine Corps Aviation: The Early Years 1912–1940*. Washington, D.C.: History and Museums Division, Headquarters, U.S. Marine Corps, 1977.

Jones, Sir Henry A. *The War in the Air*, vol. 2. Oxford: Clarendon Press, 1928.

———. *The War in the Air*, vol. 5. Oxford: Clarendon Press, 1935

————. *The War in the Air*, vol. 6. Oxford: Clarendon Press, 1937.

————. *The War in the Air*, Appendices. Oxford: Clarendon Press, 1937.

Jones, Neville. *The Origins of Strategic Bombing*. London: William Kimber, 1973.

Klachko, Mary. "William Shepherd Benson: Naval General Staff American Style." In James Bradford, ed., *Admirals of the New Steel Navy*. Annapolis, Md.: Naval Institute Press, 2012.

Klachko, Mary, with David Trask. *Admiral William Shepherd Benson, First Chief of Naval Operations*. Annapolis, Md.: Naval Institute Press, 1987.

Knappen, Theodore M. *Wings of War: An Account of the Important Contributions of the United States to Aircraft Inventions, Engineering, Development and Production During the World War*. New York: G. P. Putnam Sons, 1920.

Lake, Jon. *The Great Book of Bombers: The World's Most Important Bombers from World War I to the Present Day*. St Paul, Minn.: MBI, 2002.

Lea, John. *Reggie: The Life of Air Vice Marshal R L G Marix CB DSO*. Edinburgh: Pentland Press, 1994.

Levine, Isaac D. *Mitchell: Pioneer of Air Power*. New York: Duel, Sloane and Pearce, 1943. Also printed in London, England, by Peter Davies under the title *Flying Crusader: The Story of General William Mitchell, Pioneer of Air Power*.

Malandrino, Greg. "Alfred Austell Cunningham: Father of Marine Corps Aviation." *Over the Front* 14, no. 2 (Summer 1999): 358–67.

Marder, Arthur J. *From the Dreadnought to Scapa Flow: The Royal Navy in the Fisher Era, 1904–1919*, vol. 5: *Victory and Aftermath (January 1918–June 1919)*. London: Oxford University Press, 1970

Martel, René. *French Strategic and Tactical Bombardment Forces of World War I*. Lanham, Md.: Scarecrow Press, 2007.

Maurer, Maurer. *Aviation in the U.S. Army 1919–1939*. Washington, D.C.: Office of Air Force History, 1987.

————. ed. *The United States Air Service in World War I*, 4 vols. Washington, D.C.: The Office of Air Force History, 1978.

Meilinger, Philip S. "Giulio Douhet and the Origins of Airpower Theory." *The Paths of Heaven: The Evolution of Airpower Theory*. Maxwell AFB, Ala.: Air University Press, 1997.

Mersky, Peter. *U.S. Marine Corps Aviation, 1912 to the Present*, 3rd ed. Baltimore: Nautical and Aviation Publishing Company of America, 1997.

————. *U.S. Marine Corps Aviation Since 1912*, 4th ed. Annapolis, Md.: Naval Institute Press, 2009.

Miller, Roger G. *Billy Mitchell: Stormy Petrel of the Air*. Washington, D.C.: Office of Air Force History, 2004.

Mitchell, William. *Memoirs of World War I: "From Start to Finish of Our Greatest War."* New York: Random House, 1962.

Morris, Michael. "Combat Effectiveness: United States Marine Corps Aviation in the First World War." *Over the Front* 12, no. 3 (Autumn 1992): 232–8.

Morrow, John H., Jr. *The Great War in the Air: Military Aviation from 1909 to 1921.* Washington, D.C.: Smithsonian Institution Press, 1993.

Morton, John F. *Mustin—A Naval Family of the Twentieth Century.* Annapolis, Md.: Naval Institute Press, 2003.

Moseley, George. *Extracts from the Letters of George Clark Moseley during the Period of the Great War.* Chicago: Private Printer, c. 1923.

O'Connor, Mike. *Airfields and Airmen.* London: Leo Cooper, 2001.

Owers, Colin. "Handley Page 0/400." *Cross & Cockade International* 23, no. 3 (Autumn 1992): 120–30.

———. "Handley Page Trainee Pilot: The Experiences of Sir Laurence Hartnett in the RNAS and RAF." *Cross & Cockade International* 23, no. 3 (Autumn 1992): 113–20.

———. "Killingholme Diary," Parts I and II. *Cross & Cockade International* 35, nos. 3 and 4 (Autumn, Winter 2004): 169–95, 211–32.

Paine, Ralph. *The First Yale Unit: A Story of Naval Aviation 1916–1919*, 2 vols. Cambridge, Mass.: Riverside Press, 1925.

Patrick, Mason M. "Final Report of Chief of the Air Service A.E.F. to the Commander in Chief American Expeditionary Forces." *Air Service Information Circular*, vol. II, n180 (February 15, 1921).

Peason, Bob. "'. . . More Than Would Be Reasonably Anticipated,' the Story of No. 3 Wing Royal Naval Air Service." *Over the Front, the League of WWI Aviation Historians.* www.overthefront.com/WWI-Aviation-No-3-Wing-Royal-Naval-Air-Service-p1.php.

Pollard, Bridget. "Royal Naval Air Service in Antwerp, September–October 1914." British Commission for Military History website, www.bcmh.or.uk, 2003

Potter, Andrew E. "The 'Fiat' and the 'Caproni'." In Bludworth, *Battle of Eastleigh, England, USNAF*, pp. 96–8.

Rainey, James C., ed. in chief. *2003 Logistics Dimensions.* Maxwell, Ala.: Air Force Logistics Agency, 2003.

Reynolds, Clark G. *Admiral John H. Towers.* Annapolis, Md.: Naval Institute Press, 1991.

Robertson, F. A. de V. "No. 207 (Bomber) Squadron." *Flight*, October 12, 1933, 1022–34.

Roskill, Stephen W. *Documents Relating to the Naval Air Service*, vol. 1: *1908–1918.* London: Navy Records Society, 1969.

Rossano, Geoffrey L. "Doing Their Duty Side by Side: Allied–American Aviation Cooperation in World War I." Master's thesis, UNC–Chapel Hill, 1974.

———. *Hero of the Angry Sky: The World War I Diary and Letters of David S. Ingalls, America's First Naval Ace.* Athens: Ohio University Press, 2013.

———. *The Price of Honor: The World War One Letters of Naval Aviator Kenneth MacLeish.* Annapolis, Md.: Naval Institute Press, 1991.

———. *Stalking the U-boat: U.S. Naval Aviation in Europe During World War I.* Gainesville: University of Florida Press, 2010.

Sheely, Irving E., and Lawrence D. Sheely. *Sailor of the Air: The 1917–1919 Letters & Diary of USN CMM/A Irving Edward Sheely.* Tuscaloosa: University of Alabama Press, 1993.

Sherrod, Robert. *History of Marine Corps Aviation in World War II*, 2nd ed. New York: Presidio Press, 1980.

Shirley, Noel. "Capt. LaGuardia, Caproni Bombers, and the U.S. Navy." *Cross & Cockade International* 25, no. 2 (Summer 1984): 125–40.

———. "John Lansing Callan—Naval Aviation Pioneer." *Over the Front* 2, no. 4 (Winter 1987): 356–70.

Sims, William S. *The Victory at Sea.* Garden City, N.Y.: Doubleday, Page & Co. 1920.

Sitz, W. H. *A History of Naval Aviation.* Washington, D.C.: GPO, 1930.

Still, William N. *The Crisis at Sea: The United States Navy in European Waters in World War I.* Gainesville: University Press of Florida, 2006.

Stoff, Joshua. *The Aviation Heritage of Long Island.* Hempstead, N.Y.: The Cradle of Aviation Museum, 2001.

Swanborough, Gordon, and Peter M. Bowers. *United States Navy Aircraft Since 1911.* Annapolis, Md.: Naval Institute Press, 1990.

Sweetser, Arthur. *The American Air Service: A Record of Its Problems, Its Difficulties, Its Failures and Its Final Achievements.* New York: D. Appleton, 1919.

Taussig, Joseph K. *The Queenstown Patrol, 1917: The Diary of Commander Joseph Knefler Taussig, U.S. Navy.* Newport, R.I.: Naval War College Press, 1966.

Thetford, Owen. *British Naval Aircraft since 1912*, 6th ed. Annapolis, Md.: Naval Institute Press, 1991.

Trask, David. *Captains and Cabinets: Anglo-American Naval Relations, 1917–1918.* Columbia: University of Missouri Press, 1972.

Tucker, Ian, ed. *The European Powers in the First World War: An Encyclopedia.* New York: Garland, 1996.

Turnbull, Archibald D., and Clifford L. Lord. *History of Naval Aviation.* New Haven, Conn.: Yale University Press, 1949.

U.S. Navy, Office of Naval Records and Library. *The American Naval Planning Section, London.* Washington: GPO, 1923.

Van Deurs, George. *Wings for the Fleet.* Annapolis, Md.: Naval Institute Press, 1966.

Wait, Adam, and Noel Shirley. "'Devil Dog' Sam Richards." *Over the Front* 7, no. 3 (Autumn 1992): 195–211.

White, Robert P. *Mason Patrick and the Fight for Air Service Independence.* Washington, D.C.: Smithsonian Institution Press, 2001.

Whitehouse, Arch. *The Zeppelin Fighters.* Garden City, N.Y.: Doubleday, 1966.

Williams, George K. *Biplanes and Bombsights: British Bombing in World War I.* Maxwell Air Force Base, Ala.: Air University Press, 1999.

———. "'The Shank of the Drill': Americans and Strategical Aviation in the Great War." *The Journal of Strategic Studies* 19, no. 3 (September 1996): 381–431.

Willis, Ron L., and Thomas Carmichael. *United States Navy Wings of Gold: From 1917 to the Present.* Atglen, Pa.: Schiffer, 1995.

Wise, Sydney. "The Royal Air Force and the Origins of Strategic Bombing." *Men at War: Politics and Innovation in the Twentieth Century.* Chicago: Precedent, 1982.

Woodhouse, Henry. *Textbook of Military Aeronautics.* New York: Century, 1918.

Wortman, Marc. *The Millionaires' Unit.* New York: Public Affairs, 2006.

Wright, Peter. "Dunkerque Days and Nights." *Cross & Cockade International* 23, no. 2 (Summer 1992): 131–44.

Wright, Reginal W. *Contact.* Washington, D.C.: Naval Aviator Register, c. 1967.

Dissertations, Papers, and Other Unpublished Material

Lauderbaugh, Lt. Col. George M. "The Air Battle of St. Mihiel: Air Campaign Planning Process Background Paper." http://www.au.af.mil/au/awc/awcgate/ww1/stmihiel/stmihiel.htm#before.

Lord, Clifford L. "The History of United States Naval Aviation, 1898–1939." Clifford L. Lord Papers, Operational Archives, Naval History and Historical Command, Washington, D.C. Note: there is also a "Draft" version of this work titled "Naval Aviation History." The 1898–1939 work is the multivolume typescript series with the full title. The draft is a separate, partial, and probably earlier version with completely different pagination and often different/supplemental material. The two complement each other, but are not the same work.

"Organization of Military Aeronautics 1907–1935." Army Air Forces Historical Study No. 25.

Parker, John D. "The Early Development of United States Air Doctrine." Air University Report No. 6024, Maxwell Air Force Base, Ala., 1976

Taschner, Maj. Michail J. "Examples of Airmindedness From America's First Operation Air Campaign: The St. Miheil Offensive, 1918." Research paper presented to the Research Department, Air Command and Staff College, March 1997.

Periodicals

Aerial Age Weekly
Flight
International Military Digest Annual
Register of Commissioned and Warrant Officers of the United States Navy and Marine Corps
Technology Review (Massachusetts Institute of Technology)

Index

Great White Fleet, 176
Green Balls: The Adventures of a Night Bomber, 86
Grey, Spenser D. A., 43, 78, 207; attacks LZ39, 11–12; bombing of airship sheds, 8; Cone's advisor, 139; commands No. 5 Wing, 13; detached to Italy, 131; importance to NBG, 120, 122; inspects airfields, 140; instructs and inspires USN, 2; Lovett gives credit to, 78; and NBG initiative, 121; night training, 126; posted to France, 11; provides data to support NBG structure, 115; and U.S. bombing program, 62; and USN bombing policy, 75; visits St. Inglevert, 176. *See also* photo section
Griffin, Virgil C. "Squash," 46, 66
Guggenheim, Harry, 117, 118, 119, 131, 140: commission studying Adriatic, 207; negotiates agreement to purchase Caproni bombers, 134–35; on strategic targeting, 195

Haig, Douglas, 19, 56
Hale bomb, 9, 11, 12
Hall, Weston "Bert" 155–56
Handley Page, Sir Frederick, 170
Handley Page 0/100, 15, 59; bomb load, 100; characteristics, 88; first flight, 170; flight characteristics, 128; Lovett' flies, 83, 85; NBG requirements, 91; negotiations for assembly from U.S. supplied parts, 109–110; production U.S., 103–4, 109–11; superiority of, 199
Handley Page 0/400, 171
Handley Page V/1500, 197, 199

Hanrahan, David C., 118–120, 130, 141; on British influence, 191; called the Iron Duke, 118; gathers staff, 172; inspects St. Inglevert, 176; value of 214 Squadron, 171; visits La Fresne, 177; visits St. Inglevert, 176. *See also* photo section
Hanriot-Dupont floatplanes, 68
Hanriot-Dupont scout, 48
Harbord, James G., 53
Harnsworth, Arthur, Lord Northcliffe, 25
Harris, Stephen, 3
Hartle, Orrin, 173
Hartnett, Laurence, 128
Harts, William W., 116
Harvest Moon raids, 59
Haviland, Willis T., 155
Hazelhurst, Wilson L., 147–49
Hazelhurst Field (USAAS), 99, 147
Heldreth, Alonzo, 47
Henderson, Sir David, 17–18, 57
Hickman, Albert, 203
Hicks, Frederick C., 176
Hicks Field, 99
Higham, Robin, 3
Hispano-Suiza, 157
Holley, I. B., Jr., 60, 102
Hotel St. Antoine, 9
House, Edward M., 26
Huey, Sidney, 166, 67
Hull, Carl, 66
Hunsaker, Jerome, 205
Huntington (CA-5), 27

Ideas and Weapons, 102
Independent Force (RAF), 196
Ingalls, David S., 125, 145, 165; becomes Ace, 178;

About the Authors

Dr. Geoffrey Rossano, born in Brooklyn, New York, received his B.A. in history from Tufts University and was awarded his M.A. and Ph.D. degrees from the University of North Carolina. Dr. Rossano began researching military aviation in 1968, and his master's thesis explored American-Allied aviation cooperation during the Great War. Subsequently, he edited and annotated the letters of Kenneth MacLeish, published by Naval Institute Press as *The Price of Honor*. A Vice Admiral Edwin B. Hooper research grant from the Naval Historical Center led to his study of naval aviation activities in Europe during World War I, *Stalking the U-Boat*. He recently published *Hero of the Angry Sky*, the annotated letters and papers of David Ingalls, naval aviation's first ace. Rossano was the 2010 recipient of the Roosevelt Prize in Naval History and in 2013 received of the Admiral Arthur W. Radford Award for Excellence in Naval Aviation History and Literature.

Thomas Wildenberg is an independent historian/scholar specializing in the development of naval aviation and technological innovation in the Navy. He has written extensively about the U.S. Navy during the interwar period. His articles have appeared in several scholarly journals including the *Journal of Military History, American Neptune,* and U.S. Naval Institute *Proceedings.* He is also the author of five books on U.S. naval history covering such varied topics as replenishment at sea, the development of dive bombing and the history of the torpedo in the U. S. Navy. His most recent work published by the Naval Institute Press is *Billy Mitchell's War with the Navy: The Interwar Rivalry Over Air Power.*

Mr. Wildenberg served as a Ramsey Fellow at the National Air and Space Museum in 1999-2000. He is a recipient of Arthur W. Radford Award for Excellence in Naval Aviation History (2012), the Surface Navy Association Literary Award (2005), two John Laymen Awards from the North American Society for Oceanic History for best naval history (2013) and best biography (2003). He received the Air Force Historical Foundation's award for the best article in the 2009 volume of *Air Power History,* was awarded an honorable mention in the Ernest J. Eller Prize in Naval History (1994), and received the Edward S. Miller Naval War College Research Fellowship (1998).

The **Naval Institute Press** is the book-publishing arm of the U.S. Naval Institute, a private, nonprofit, membership society for sea service professionals and others who share an interest in naval and maritime affairs. Established in 1873 at the U.S. Naval Academy in Annapolis, Maryland, where its offices remain today, the Naval Institute has members worldwide.

Members of the Naval Institute support the education programs of the society and receive the influential monthly magazine *Proceedings* or the colorful bimonthly magazine *Naval History* and discounts on fine nautical prints and on ship and aircraft photos. They also have access to the transcripts of the Institute's Oral History Program and get discounted admission to any of the Institute-sponsored seminars offered around the country.

The Naval Institute's book-publishing program, begun in 1898 with basic guides to naval practices, has broadened its scope to include books of more general interest. Now the Naval Institute Press publishes about seventy titles each year, ranging from how-to books on boating and navigation to battle histories, biographies, ship and aircraft guides, and novels. Institute members receive significant discounts on the Press's more than eight hundred books in print.

Full-time students are eligible for special half-price membership rates. Life memberships are also available.

For a free catalog describing Naval Institute Press books currently available, and for further information about joining the U.S. Naval Institute, please write to:

Member Services
U.S. Naval Institute
291 Wood Road
Annapolis, MD 21402-5034
Telephone: (800) 233-8764
Fax: (410) 571-1703
Web address: www.usni.org